Hybrid Microelectronic Circuits
The Thick Film

RICHARD A. RIKOSKI

The Moore School of Electrical Engineering
University of Pennsylvania

A WILEY-INTERSCIENCE PUBLICATION

JOHN WILEY & SONS, New York • London • Sydney • Toronto

Library of Congress Cataloging in Publication Data

Rikoski, Richard A
    Hybrid microelectronic circuits: the thick film.

    "A Wiley-Interscience publication."
    Includes bibliographical references.
    1. Microelectronics. I. Title.

TK7874.R55   621.381'7   73–7928
ISBN 0-471-72200-6

Printed in the United States of America

10 9 8 7 6 5 4 3 2 1

TO

EVERARD M. WILLIAMS and FREDERICK J. YOUNG

HYBRID MICROELECTRONIC CIRCUITS

FOREWORD

Over the years organizations such as the IEEE have given much serious thought and considerable time and effort to the manner in which they can best serve the educational needs of practicing engineers. More specifically, in recent years they have devoted much of their energy toward evolving new programs of continuing education. This book represents one important step in this evolutionary process inasmuch as its writing was commissioned by the Educational Activities Board of the IEEE as part of a developmental program in self-study. This book, therefore, can serve the reader in two significant ways: as a text to which he can turn for authoritative information on hybrid integrated circuit technology in the manner in which he is normally accustomed to use such a text, and as part of a complete self-study package on the topic available from the IEEE.

The Educational Activities Board of the IEEE commends the author, Professor Richard A. Rikoski of The Moore School of Electrical Engineering, University of Pennsylvania, for his dedicated effort in not only writing this text, but also integrating it into a practical unit of self-study material of use to a wide range of engineers in both industry and academia.

Glen Wade, Chairman
1970–1972 IEEE Educational Activities Board

PREFACE

Thick film technology enjoyed a brief period of acceptance in the early 1960s but fell into disuse with advances in monolithic technology. However, the thick film hybrid steadily regained popularity after its advantages over monolithic circuits became better known and appreciated.

The thick film performs better in high-frequency, high-voltage, and high-power applications. Moreover, thick film hybrid microcircuits are relatively easy to design and construct and are less expensive in terms of initial equipment, circuit development, and production cost. Thick film hybrids are especially suited to small manufacturing runs and can easily accommodate last-minute circuit design changes. Although it has a smaller function density, the appearance of the thick film package is similar to that of the monolithic. Today thick film hybrid modules are found in a variety of complex electronic systems, such as large-scale digital computers and the majority of color television receivers marketed in the United States.

Interest in the thick film hybrid has grown steadily since its rebirth. Two professional societies, The Institute of Electrical and Electronic Engineers and the International Society for Hybrid Microelectronics, fostered that growth. I hope that this monograph will introduce the solid state designer who is competent in the physical realization of solid state discrete component electronics to the thick film art that will serve him well throughout this decade.

Chapter 1 is a survey of microelectronics, including monolithics, thick and thin films, and a quick look at the hybrid production process. Subsequent chapters amplify such elements of these topics as substrates, film materials, chip components, bonding, screening, firing, resistor trimming, packaging, and quality control testing. For example, Chapter 5 gives substrate layout and circuit design suggestions, whereas Chapter 7 is devoted to the subject of interacting with vendors. The Appendix contains a bibliography of texts, proceedings, and journals and an abridged list of material and circuit producing companies that provide further information on particular thick film topics.

I express my gratitude to the Educational Activities Board of the Institute of Electrical and Electronics Engineers, Inc., especially to IEEE Director of Educational Services, John M. Kinn; to Joseph E. Casey and Emma White for their continued support; and to Judy Lehman for her excellent typing of the manuscript.

I deeply appreciate the interest and cooperation of the members of 112 industrial concerns and laboratories involved with this technology who generously provided much of the information, suggestions, and illustrations that make up this book.

RICHARD A. RIKOSKI

Philadelphia, Pennsylvania
February 1973

CONTENTS

Chapter 2 Substrates

Chapter 3 Screened and Fired Conductive, Resistive and Dielectric Films

Chapter 4   Chip Components

Chapter 5   Design Comments

Chapter 6   Hybrid Microelectronic Circuit Production

Chapter 7 The Hybrid Microcircuit Vendor

Appendix

ILLUSTRATIONS AND CREDITS

Chapter 1

INTRODUCTION

Hybrid microcircuit technology is a blend of fifth-century screen-process printing arts and twentieth-century electronic engineering science.

Silk screening was a widely practiced art form throughout the United States following World War I. Today it is used primarily in the production of custom printing and poster art.

Twentieth-century electronics has progressed from early coherers and other electromechanical devices through vacuum tube and transistor development to the microelectronic circuits of the 1970s: the monolithic, the hybrid thin film, and—the subject of this book—the hybrid thick film.

This chapter introduces hybrid thick film technology and presents an outline of the book.

SOME BASIC DEFINITIONS

MICROELECTRONICS. Microelectronics is the art and science of electronic circuits and systems realized from extremely small electronic devices.

1

MONOLITHIC INTEGRATED CIRCUIT. The physical realization of a number of electronic circuit elements inseparably associated within a continuous body of material (usually semiconductor) to perform electrical circuit functions.

HYBRID MICROELECTRONIC CIRCUIT. The physical realization of an electronic circuit from two or more discrete circuit elements, such as monolithic integrated circuits, transistors, diodes, resistors, capacitors, transformers, and conductors. These elements usually are mounted on an insulating substrate upon which conducting lines, resistors, capacitors, and inductors have been deposited to produce circuit functions similar to, but often with advantages over, monolithic integrated circuits.

HYBRID THIN FILM CIRCUIT. A hybrid microelectronic circuit whose substrate films have been deposited by gaseous or vapor means.

HYBRID THICK FILM CIRCUIT. A hybrid microelectronic circuit with substrate depositions implemented by screening and firing layers of high-viscosity pastes composed of noble metal powders, vitreous binders, and an organic vehicle.

MAJOR ADVANTAGES OF MICROCIRCUITS

Microcircuits—whether monolithic, thick film hybrid, or thin film hybrid—can save production costs while delivering miniaturized, highly reliable circuits. Production line savings accrue since parts procurement, handling, and inventories are simplified by eliminating large numbers of discrete components. Miniaturization reduces size, weight, and usually power supply requirements as well (1).

Microcircuits are highly reliable since the entire package is sealed and leads are virtually eliminated by screening or depositing conductors. In addition, quality control is easier to perform on a microcircuit than on a printed circuit board, since one circuit rather than scores of components is being tested. Test data, too, are usually more meaningful for one component than for many.

Resistors produced by screening and trimming or by diffusion can be more accurate and track better than those purchased as components.

Advantages apply to the entire system as well. For instance, digital circuit speed and amplifier high-frequency performance and stability are enhanced since lead length capacitance is sharply reduced. Elements can be positioned with little regard for interconnections. Increased circuit density, which accompanies miniaturization, reduces the required number of printed circuit boards and connectors per system (2).

SILICON MONOLITHIC INTEGRATED CIRCUITS

Monolithics (3) are formed by successively diffusing dopants into a silicon substrate forming p and n regions that comprise bipolar and field effect transistors, diodes, and passive components. These are interconnected through the substrate itself or by subsequent vapor deposited metallizations.

Among the general characteristics of monolithic integrated circuits are the following:

- They combine active and passive elements in a single "monolithic structure."
- There is high component density within a standard package.
- Since they are designed for mass production, they are economical to produce in large quantity.
- They are poor for low-volume production because of expensive tooling.
- Development costs of a new silicon monolithic circuit range from $2000 to $35,000.
- Expensive laboratory facilities, including clean rooms, are required for production.
- A monolithic circuit available "off the shelf" from a vendor, which meets all circuit requirements, is probably more economical than design and production of new hybrid.
- Integrated circuits usually are optimized with respect to active devices, resulting in wider tolerances on resistors and capacitors.
- Designs using monolithics often are complicated by the presence of unwanted pn junctions with the substrate.
- Delivery time from vendors generally is 11 to 26 weeks.

THIN FILM HYBRID MICROCIRCUITS

The thin film hybrid (4–9) is fabricated by vapor depositing conductors and components onto a passive substrate. Add-on devices specially chosen for best circuit performance—including monolithics, junction and field effect transistors, and various passive elements—are then attached to the substrate. The entire circuit subsequently is encased within an epoxy dip or a hermetically sealed package.

Among the general characteristics of thin film hybrids are the following:

- Specially chosen diodes, junction and field effect transistors, monolithics, capacitors, resistors, and inductors can be included within the microcircuit.
- These hybrids are excellent for very small circuits such as $\frac{1}{8} \times \frac{1}{4}$ in. substrates.
- The process takes place in a vacuum chamber. Several steps, each requiring 10 to 30 minutes for evacuation, may be necessary.
- Thin films (less than 2000 Å (1/125 mil thick) are deposited, atom by atom, upon a glass or glazed alumina substrate.
- The process generally used is vacuum evaporation: conducting material is heated in a vacuum, atoms of this material vaporize, they then diffuse toward the cold substrate and condense upon it.
- Sputtering, polymerization, vapor plating, and anodization or other techniques are used for producing thin films.
- Materials used are copper and gold conductors, silicon monoxide insulating layers, and tantalum, nichrome, and tin oxide resistors.
- An exact conductor pattern is obtained by masking or photolithographic etching.
- Thin film properties are very sensitive to deposition parameters such as substrate temperatures and surface conditions, diffusion chamber atmosphere, and sputter source composition and temperature.
- These hybrids require very smooth substrates.

- Resistance can be measured throughout diffusion resulting in 5% tolerances without trimming and TCR of 50 ppm/°C or better.
- Sheet resistivities are 25 to 10,000 $\Omega/\square$.
- Fine conductor line widths (1 mil) are required.
- The initial investment for production equipment is high (typically $500,000).
- Development costs usually are higher than those of thick films.
- Delivery time from vendors is generally 6 to 10 weeks.

THICK FILM HYBRID MICROCIRCUITS

Thick film hybrid microcircuits, (10–14; Figures 1.1 and 1.2) are also produced by depositing conductor and passive-component films onto a passive substrate. However, the thick film technique uses a paste screening (similar to silk screening) and firing process that is simpler than vapor deposition. Again, add-on components, chosen

Figure 1.1. Typical hybrid thick film circuits. Courtesy of Electro-Science Laboratories, Inc.

Figure 1.2. Thick film hybrid circuit. Courtesy of Intersil, Incorporated.

for best circuit compatibility, are bonded to the substrate with the entire circuit suitably packaged in plastic, ceramic, or metal.

The general characteristics of thick film hybrids include the following:

- A rougher alumina substrate is used; the conductor, resistor, and dielectric pastes are screened and fired, with semiconductors, and passive elements are later bonded onto this substrate.

- Thick films are pastes containing fine powders, which are "silk" screened onto a substrate and then furnace fired at high temperature.

- Typical film thickness is 0.5 to 1.5 mils.

- The manufacturing process used is simpler than that required for monolithics or thin films.

- The initial capital investment is low $100,000 to $125,000 (minimum $25,000) for a highest quality prototype facility. Production capability costs from $100,000 to $500,000 including training.

- Tooling for a new circuit cost $500 to $1500; this is the cheapest way to get into microelectronics.

- High-frequency performance is excellent.

- This hybrid can be used at much higher power and voltage levels than can a monolithic circuit.

- Conversion from breadboard to hybrid thick film circuit is easy.

- High yields are possible.

- External appearance is similar to a monolithic circuit; it can be hermetically sealed.

- The crossover capability is the same as that of a printed circuit board but not as expensive. Four or five layers of conducting, resisting, and insulating films are screenable.

- Delivery time from vendors is 6 to 10 weeks.

THE CASE FOR SCREENED AND FIRED THICK FILM HYBRIDS

Having considered the merits of the three major microcircuits, we recapitulate the advantages that make the hybrid thick film technique worthy of attention:

1. Hybrid thick films are the least expensive way of getting into microcircuit production—especially when small quantities of many different types of microcircuits are required.
2. The thick film technique, because of its inherently low parasitic capacitance, provides excellent performance at high frequencies.
3. Hybrids are most useful in miniaturizing high-voltage and high-power circuits.
4. Cost advantages accrue from simplifying parts inventories and procurement, simplifying test and assembly procedures, and reducing the amount of interconnection hardware.
5. Hybrids allow greater design and production flexibility than monolithics by combining the advantages of different discrete and monolithic technologies while maintaining the outward package appearance of a large monolithic.
6. The hybrid technique, in contrast to the monolithic, permits optimizing the selection of both active and passive elements.
7. The thick film process is relatively simple and is capable of excellent yields.
8. Direct circuit translation from breadboard to microcircuit, as well as subsequent circuit changes, are accomplished with relative ease.

In conclusion, the hybrid thick film combines cost, design, and flexibility advantages over the monolithic and thin film with the penalty of loss in component density. Considering the current state of the art, high-power, high-frequency, and high-voltage circuits are best miniaturized by the hybrid.

THICK FILM PRODUCTION

Thick film technology centers around the application of high viscosity pastes onto ceramic substrates by printing through a fine mesh stainless steel screen or mask. Firing the screened substrate at about

1000°C causes the paste to adhere to the substrate as a permanent 1-mil-thick film. The screening and firing process can be repeated to add layers of conductive, dielectric, or resistive films. Various trimming techniques are used to obtain resistance and capacitance values with tolerances better than 5%.

Semiconductor (Figure 1.3) and passive component chips are

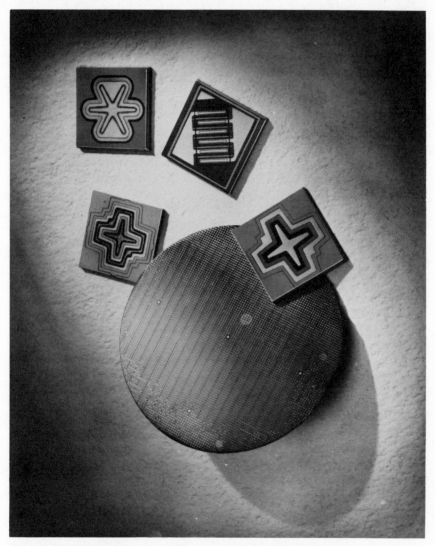

Figure 1.3. Semiconductor chips for thick film hybrid applications. Courtesy of Motorola Semiconductor Products, Inc.

next attached to the substrate and the circuit is encased in plastic, ceramic, or metal. The resultant thick film hybrid microelectronic circuit is then tested and delivered for application.

SUBSTRATES

The substrate provides physical support for components, electrical insulation between conductors, lead tie points, and a means of heat transfer.

Substrates must be relatively smooth, flat, and capable of withstanding, without chemical decomposition, the sustained heat of a 1000°C thick film firing cycle.

Common thick film substrate materials are alumina, beryllia, and barium titanate.

Further details are given in Chapter 2.

DESIGN, CIRCUIT LAYOUT, ARTWORK

The design phase translates a conventional breadboarded design to a microcircuit form. The designer, who considers performance requirements of the final circuit in the light of his knowledge of the requisites and limitations of the hybrid thick film process, specifies components, tolerances, package type, and substrate layout. He accounts for substrate "real estate" limitations, component power dissipations, and the placement of screened conductors, lands, and crossovers. He specifies screened resistor design and trim. Further design details are noted in Chapter 5.

Master drawings of conductor and resistor patterns, 20 times larger than actual circuit size are used to outline Rubylith masters. These in turn are used to photoetch mesh screens through which the thick film will be printed on the substrate.

THICK FILM SCREENING

In the screening or printing process a rubberlike squeegee forces thick film, conductive, resistive, or dielectric paste through a screen that has been photoetched to the shape of the circuit.

Commonly used conductive or resistive inks are platinum-gold, palladium-gold, gold, palladium-silver, and silver. Dielectric pastes are barium titanate, silicon dioxide, aluminum oxide, tantalum pentoxide or glass-ceramic. Proprietary blends are also used. Paste selection is detailed in Chapter 3.

FIRING

After screening the substrate is passed through furnace where temperatures of 1000°C bake the paste into position on the substrate.

Each substrate is usually subjected to several screening and firing steps. Conductors are screened and fired first; then capacitor dielectrics, capacitor top plates, and finally thick film resistors are enplaced. A final encapsulating glaze coat then is screened and fired.

Each step generally requires its own screening and firing cycle; each firing is carried out at successively lower temperature.

ADJUSTING VALUES

The screening and firing process is sufficiently accurate to produce resistors and capacitors with ±20% tolerance. Closer tolerances may be obtained by adjusting component values or "trimming" after firing.

One common means of trimming is "Airbrasion," in which a finely directed abrasive particle airblast removes resistor or capacitor material. Measuring probes attached to the resistor and a feedback control system automatically stop the process when correct tolerance is achieved.

Other trimming techniques, which include the use of lasers, ultrasonic chisels, and electrical or thermal pulses, are described in Chapter 3.

CHIP COMPONENTS

It is the chip or add-on component that makes a hybrid a hybrid. There are many types of active elements that can be easily inserted into these circuits.

Configurations include simple chips with conducting lands wire bonded to the substrate as well as a variety of leadless types including flip chips, beam leads, and ceramic flip chips. These are bonded by simple soldering or welding, thermocompression bonding, ultrasonic bonding, adhesives, evaporated connections, and other methods. Passive chips are added to the substrate in place of screened and fired components when there is design advantage in doing so.

Chip resistors are sometimes used to eliminate separate screening and firing cycles when many different resistor inks would be required to achieve a wide resistance spread in the circuit design. Chip resistors also allow quick implementing of circuit design changes.

Chip capacitors, which deliver much higher capacitance per unit volume, are most frequently used to provide capacitance. The use of chips also eliminates the multiple screening and firing stages required to produce thick film capacitors.

Microcircuit inductors and their elimination or inclusion within the hybrid are considered in Chapter 4, as are chip resistors, capacitors, and active elements.

PACKAGING

The package that encases the hybrid provides mechanical support for the circuit, isolates it from hostile environments, and also serves as a heat sink. Common packages are TO5 and TO8 cans, flat packs, and epoxy shells.

Artwork, screening, firing, and packaging are treated in Chapter 6.

QUALITY CONTROL

The opportunity to monitor physical and electrical properties exists throughout the hybrid production process, from the initial procurement of materials, inks, substrates, and chip components through screening, firing, and component trimming, to final measurements which evaluate the fabricated and packaged hybrid.

Several companies make die matrix test stations. These contain microscopes, probes, and positioners for testing and manipulating chips and partially finished hybrids. Tensile and shear tests can be made of bonded leads, and stress-strain gauges are available to test substrate tolerance and warpage and squeegee hardness; they also screen tolerance and emulsion thickness.

Complete test systems are available to measure a variety of electrical parameters manually, semiautomatically, or with full automation, depending on production requirements.

COSTS, VENDORS, AND THE QUESTION OF MAKE OR BUY

The thick film approach allows one a relatively inexpensive entry into microelectronic manufacture. Initial equipment costs can range from $25,000 to $500,000 depending on whether a minimal prototype or full production capability is required. An excellent prototype facility can be assembled for $100,000 to $125,000, exclusive of personnel costs.

Investing in a full production facility, at least initially, is not necessary because many vendors who have such capability are willing to supply custom-made hybrids at reasonable prices with short delivery times. Moreover, a quality vendor usually has an excellent engineering staff to assist in circuit design and package selection while being mindful of cost and environment factors. The vendor can usually perform a wide range of quality control and reliability testing on his finished circuits. This topic is considered in Chapter 7.

William J. MacDonald (15), President of Film Microelectronics, Inc., offers the following comments on cost:

. . . Generally when an integrated circuit is available, it'll be less costly to use than a hybrid circuit. However, if there is no standard IC available, then the hybrif circuit will be less costly if it's required in quantities that are less than 50,000 units. The cost to develop the first of an IC is approximately $50,000, $10,000 to alter an established IC. This engineering write off makes the cost of an IC high in small quantities. This is the logic behind the use of hybrid circuits in the small quantity runs . . .

. . . The list below indicates the cost centers in building a hybrid and their percentage of the total cost as we see them at FMI.

Cost Center	Percentage Total Cost
1. Passive Network	5–15%
2. Added Components	40–60%
3. Labor	10–15%
4. Package	10–20%
5. Test	10–15%

As this list indicates, the largest cost factor in the hybrid are the components added. The transistors, diodes, capacitors, IC's, etc., are the components that determine the value of the hybrid. The ability to buy these components in volume at the low price is most important. The large company that has strong buying power or the semiconductor manufacturer who can use components in a hybrid circuit with a single overhead markup has a decided advantage . . . .

. . . We find the labor content of the hybrid to be relatively small (10–15%). This fact says that we would gain very little in shipping our assembly to Hong Kong, or investing capital to eliminate labor. One should not conclude, however, that mechanization is of no value. We use it to increase our production through-put. However, we do not look to mechanization to drastically reduce our cost.

The figure that I have used for packaging (10–20%) is an average figure. Hermetically sealable flatpacks are very expensive. On the other hand, the hermetic, TO5, TO8, metal cans, etc., and plastic flatpacks are relatively inexpensive. We much prefer to use a 20¢ TO-8 can than a $\frac{1}{2}$ in. flatpack that will cost approximately $1.00 to $1.50. The need for inexpensive hermetic packages in the larger sizes is a definite need by the hybrid industry.

The cost of testing varies widely with the type of circuitry. The 10–15% that I have indicated is again an average number. There are instances where the complexity of the hybrid circuit requires testing that is as costly as the manufacture of the circuit. Most of the time, however, testing has been reduced to a less than one minute operation by the time a circuit reaches volume manufacture. More elaborate spot check testing is then used. . . .

MacDonald also pointed out that for higher production quantities, the thin film hybrid technique is competitive with the thick and could be cheaper.

Finally, an important cost that must be considered is that of the engineering, technician, and direct labor payroll of the thick film staff. A minimum of five or six people seems to be required for a prototype shop; the size of the group increases with the ambitiousness of the plant. Thus direct personnel costs, not considering burden rate, will begin at $40,000 to $50,000 per year.

THICK FILM APPLICATIONS

Circuit applications that are well suited to miniaturization by the thick film hybrid approach include the following:

1. VHF and UHF receivers and transmitters.
2. Operational amplifiers.

3. Active filters.
4. Oscillators.
5. D/A and A/D converters.
6. Power amplifiers.
7. Computer driver and logic circuitry.
8. Analog switches (FET and BJT).
9. Voltage regulators.
10. High-voltage circuits.
11. Video amplifiers.

SUMMARY

The fundamental definitions of microelectronics were presented in this chapter. The characteristics, advantages, and disadvantages of the monolithic integrated circuit and thick and thin film forms of microelectronic circuit synthesis were compared. An overview was given of the thick film art including substrates, circuit design and layout, screening, firing, resistor trimming, and chip components. Quality control, costs, vendors, and the question of making or buying the hybrid were mentioned. The pertinent comments of the president of a leading custom hybrid vending company were cited. Some applications for thick film were also given.

REFERENCES

1. J. J. Staller, "Introduction to Thick Film Hybrid Circuits," *IEEE Workshop on Thick-Film Hybrid IC Technology*, p. 1.4, March 22, 1968.
2. E. H. Melan, "Recent Developments in Thick-Film Hybrid Modules," *SCP and SST*, Vol. 10, No. 6, pp. 23–24, June 1967.
3. V. K. Mitrisin, "A Review of Process Equipment for an Integrated Circuit Facility," *SCP and SST*, Vol. 10, No. 8, pp. 53–66, August 1967.
4. P. H. Hall and N. Laegreid, "Minimum Facilities for Production of Thin-Film Hybrid Circuitry," *IEEE Conv. Technical Applications, Session* TA-1, pp. 414–415, March 1970.
5. S. B. Ruth, "Hybrids . . . Thick and Thin," *Electronic Engineer*, Vol. 28, No. 10, pp. 60–63, October 1969.
6. J. R. Rairden, "Thick and Thin Films for Electronic Applications—Materials and Processes Review," *SST*, Vol. 13, No. 1, pp. 37–41, January 1970.
7. J. A. Morton, "Strategy and Tactics for Integrated Electronics," *IEEE Spectrum*, Vol. 6, No. 6, pp. 26–33, June 1969.

8. R. E. Thun, "Thick Films or Thin," *IEEE Spectrum*, Vol. 6, No. 10, pp. 73–79, October 1969.

9. R. E. Wimberly, "Comparisons and Applications Various Types of Hybrid Circuits," *SCP and SST*, Vol. 8, No. 3, pp. 46–49, March 1965.

10. J. J. Cox, Jr., "What Is Needed to Get Started in Thick Films," *IEEE Conv. Technical Applications, Session* TA-1, p. 413, March 1970.

11. D. C. Hughes, Jr., "Tooling and Part Handling Systems for Thick Film Microcircuits," *SST*, Vol. 11, No. 6, pp. 35–41, June 1968.

12. D. T. DeCoursey, "Materials for Thick Film Technology—State of the Art," *SST*, Vol. 11, No. 6, pp. 29–34, June 1967.

13. D. Boswell, "The History and Future of Hybrid Microelectronics," *Electronic Engineering*, Vol. 42, pp. 54–58, June 1970.

14. C. Marcott, "Prototyping Microcircuits," *Electronic Products*, Vol. 10, No. 7, pp. 2–13, December 1967.

15. W. J. MacDonald, private correspondence, July 28, 1970.

Chapter 2

SUBSTRATES

The passive substrate upon which a thick film hybrid circuit is screened and fired physically supports the thick film and other mounted components, transfers heat away from resistors and active devices, and provides rigid mounting for leads that interconnect the hybrid circuit to the larger external electronic system. Additionally, the thick film substrate as a passive material insulates the conductors screened on it.

An ideal substrate then is physically rigid, has low thermal resistivity, exhibits low electrical conductivity, and possesses good thermal stability throughout the high-remperature thick film firing cycle. Good adhesion exists between the substrate and its screened films. The ideal substrate should be available at low cost, in quantity, in a variety of drilled or plain shapes.

SUBSTRATE MATERIAL PROPERTIES

Thick film substrates usually are produced from alumina ceramics, beryllium oxide, and barium titanate and occasionally from steatite, zirconia, and magnesium oxide. The final electrical properties of screened and fired films depend greatly on the physical characteristics of the substrate base material. For instance, temperature coefficient of resistance and sheet resistivity are two important film properties that depend on the expansion coefficient of the underlying substrate.

The most commonly used substrate material is 96% alumina (Al_2O_3) mixed with 4% glass. Combining high physical strength, good thermal stability, and excellent dielectric properties and allowing good adhesion of most thick film pastes, alumina is available from a variety of vendors in sizes to 4 × 5 in. In volume these substrates cost about 5¢/in.². Hole patterns are available at small extra cost. Since alumina is very hard, subsequent hole drilling by the user requires abrasives such as diamond or silicon carbide or the use of a high-power laser.

Alumina substrates are easily purchased with tolerances of $\pm\frac{1}{2}\%$ on linear dimensions and 4 mils/in. in camber. Other alumina substrates of 99.5% purity can be specified when low loss at microwave frequencies is important. Figure 2.1 is a photograph of a number of alumina substrates, with and without holes, for thick film applications.

Beryllium oxide (BeO 96 to 99%) substrates in 10-mil thicknesses are used when the need for high thermal conductivity justifies a cost increase of 10 times over the cost of alumina. Beryllia substrates offer good compatibility with most thick film pastes while producing slight changes in physical properties of deposited films.

Barium titanate substrates have higher dielectric constants than alumina and thus are used frequently where the substrate itself forms a capacitor dielectric. Compared to alumina, however, titanates are structurally weaker, show more warp and camber, and produce higher sheet resistivities from alumina-designed inks.

SUBSTRATE SIZES AND TOLERANCES

Standard size substrates should be specified whenever possible to avoid the cost and time spent in obtaining nonstandard items. Typical sizes for single circuit substrates range from $\frac{1}{8}$ × $\frac{1}{8}$ in. or $\frac{1}{8}$ in. in diameter to about 2 × 2 in. or 2 in. in diameter. Van Keuren,

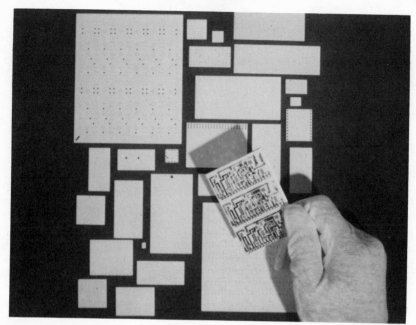

Figure 2.1. Alumina ceramic substrates for thick film circuits. Courtesy of Coors Ceramics, Inc.

for example, offers 99.5% alumina at 1 × 1, 1 × 2, and 2 × 2 in. substrates as stock items, with other sizes available upon request. Coors stocks round substrates 1.083 to 1.640 in. in diameter, square substrates 0.090 to 3.5 in. on each side, and rectangular substrates 0.052 × 1.289 and 0.093 × 0.227 to 4.000 × 4.681 in. Sizes larger than 2 × 2 in. are ordinarily used for multiple circuit applications where two or more circuits are screened onto one substrate and later separated by segmenting the substrate.

Substrates in the larger sizes can be purchased with breaklines already embossed within the substrate for easy separation of multiply screened circuits. Figure 2.2 is a photograph of alumina substrates with preembossed score lines.

Alumina substrates are commonly sized between 0.010 and 0.080 in. (±0.002 in.) thickness. The most popular single-function substrate sizes are 1 × 1 × 0.024 in. and 1 × 2 × 0.025 in. Linear dimension tolerances are usually ±1% with ±½ or better available on special order.

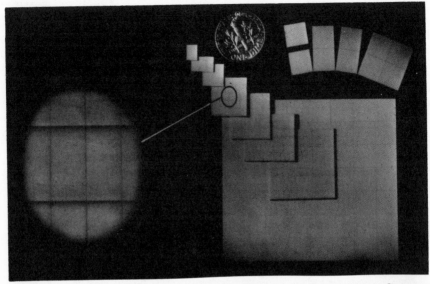

Figure 2.2.   Multicircuit alumina substrates with preembossed score lines. Courtesy of Varidyne, Inc.

Figure 2.3 is a tabulation of important physical, thermal, and electrical properties of commonly used thick film substrates.

SURFACE FINISH AND CAMBER

Since the resistance of screened and fired resistors depends on film thickness, resistor paste must be screened to a uniform height to maintain resistor tolerance. A substrate with a significant amount of camber or with an irregular surface finish will prevent this uniformity. Therefore questions of substrate flatness and finish tolerance are of great importance to the thick film manufacturer. Nonetheless, one of the advantages of thick film over thin film processing is that the thick film will tolerate a rougher substrate surface.

The standard specification on substrate camber or flatness is 0.004 in./in. Considering a 1 in.² substrate, for example, the center should not be depressed more than 0.004 in. below each edge. The interpretation of this specification is often statistical. A typical manufacturer will sample a given lot of substrates to determine the mode (most frequently occurring) thickness. This sampling is done by attempting to pass, without forcing, each substrate of a given lot between two parallel plates, spaced by a distance equal to the mode of the substrate thickness plus the camber-length product. Ground ceramic plates are used to test 96% aluminum substrates, whereas glass plates are used to prevent scratching in testing softer 99.5% alumina substrates. Camber is specified between 0.002 and 0.006 in./in. with 0.004 in./in. a good compromise between cost and quality.

Substrate roughness is also of interest. Again hybrid thick films show an advantage over thin since surface roughnesses of 20 to 40 μin. will not affect thick film properties. (Thin films require substrate smoothnesses of 1 to 10 μin. Specially processed glazed substrates with this smoothness cost 25 to 100% more than as-fired 25 μin. alumina thick film substrates.)

HOLES, FLATS, AND LANDS

Although substrates are most frequently used in rectangular form, they can be specified in disk shapes to fit can-type packages. Here a ground flat at the edge of the round substrate insures correct alignment during repeated screening and firing.

Substrates are often purchased with prepunched lands and holes to accommodate the attachment of chip components and leads. These can simplify layout, circuit assembly, and interconnection, but the design engineer must carefully judge hole and land placement to maintain adequate substrate physical strength.

Holes, indentations, alignment keys, score lines for multiple substrates, and the like, are best pressed by the substrate vendor rather than the user. Although the user can later cut, drill, or alter the substrate with diamond or Carborundum abrasives, slurry cutters, or lasers, processing by the vendor will cut both expense and work. For a special order for example, Coors can deliver several hundred substrates with 15-mil or larger holes (maximum 50 holes per piece) within a week.

Typical substrate hole dimensions are 0.020 to 0.030 in. in diameter with standard tolerances of ±10%. Tighter tolerances or holes closer than two hole diameters from an edge will increase cost substantially (1).

Substrates

	Dielectric Constant 1 MHz 25°C	Volume Resistivity 25°C (Ω cm)	Dielectric Strength (V/mil) 60 Hz	Loss Tangent (Dissipation Factor) 25°C		Loss Factor	
Alumina 95% Al₂O₃	8.9	1.9 × 10¹⁴	250	0.0002		0.0017	
Alumina 96% Al₂O₃	9.2	>10¹⁴	225 (0.250 in. thick) 500 (0.250 in. thick)	0.0010 0.0004 0.0004	1KHz 1mHz 100mHz	0.009 0.004 0.004	1KHz 1mHz 100mHz
Alumina 99.5% Al₂O₃	9.8	>10¹⁴	225 (0.250 in. thick) 550 (0.025 in. thick)	0.0002 0.0001 0.0001	1KHz 1mHz 100mHz	0.002 0.001 0.001	1KHz 1mHz 100mHz
Barium titanate (BaTiO₃)	6500	12.0	—	—		—	
Beryllia 95% BeO	6.5	>10¹⁴	—	0.0002	1mHz	0.001–1mHz	
Beryllia 99.5% BeO	7.0	10¹⁵	300 (⅛ in. disk)	0.001	1mHz	—	
Beryllia 99.9% BeO	6.7	10¹⁷	350 (⅛ in. disk)	0.0002	1mHz	—	

Figure 2.3   Characteristics of alumina, titanate,

	Density (g/cm³)	Maximum Temp. (°C)	Thermal Conductivity (cal cm/sec cm²°C)		Thermal Coefficient of Expansion per °C		Hardness Mohs Scale (Rockwell 45N)
Alumina 95% Al₂O₃	3.71	1,500	0.072	25°C	5.6×10^{-6} 6.6×10^{-6} 7.2×10^{-6}	25–100°C 25–400°C 25–700°C	9 (74)
Alumina 96% Al₂O₃	3.75	1,700	0.063 0.048 0.029	20°C 100°C 400°C	5.9×10^{-6} 7.1×10^{-6} 8.1×10^{-6}	25–200°C 25–500°C 25–1000°C	(79)
Alumina 99.5% Al₂O₃	3.90	1,750	0.075 0.065 0.028	20°C 100°C 400°C	6.0×10^{-6} 7.3×10^{-6} 7.9×10^{-6}	25–200°C 25–500°C 25–800°C	(84)
Barium titanate (BₐTᵢO₃	5.4	1,500	0.003	25°C	8.1×10^{-6}		—
Beryllia 95% BeO	2.84	1,800	0.50 0.24	25°C 300°C	8.1×10^{-6}	25–722°C	—
Beryllia 99.5% BeO	2.95– 3.00	2,200	0.63–0.67		9×10^{-6}	25–1,000°C	9
Beryllia 99.9% BeO	2.94– 2.99	2,450	0.65–0.67		9×10^{-6}	25–1,000°C	9

and beryllia thick film substrate materials.

SUMMARY

The material properties of alumina, barium titanate, and beryllia substrates were compared. Alumina was presented as the common substrate material. Like the others, it is available in a variety of shapes and sizes, with and without holes, from many vendors.

REFERENCE

1. C. E. Nordquist, "Design Criteria for Ceramic Substrates," *1968 Hybrid Microelectronics Symposium*, pp. 405–416.

Chapter 3

SCREENED AND FIRED CONDUCTIVE, RESISTIVE AND DIELECTRIC FILMS

The hybrid thick film microcircuit differs from monolithic and thin film circuits in that conductor metallizations are produced not by high-vacuum vapor depositions but by a silk screen-like process. Glass-bound metallic oxide or organometallic compounds held within an organic vehicle are screened upon a substrate by a moving rubber squeegee blade. Substrates with deposited thick films are then drawn through a firing furnace where the organic vehicle is evaporated, the thick film is aged, and the glass frit is fired at high temperature. The result is a stable, glaze-coated 0.2- to 2-mil thick film. Thick film pastes are also used to produce screened and fired resistors and capacitors. Figure 3.1 shows the consistency of a typical thick film ink.

SCREENED AND FIRED CONDUCTORS

A conductor ink formulation is a blend of finely divided metal and glass powders that are suspended in an organic vehicle much like paint thinner, which allows the ink to be screened. After screen

Figure 3.1. The consistency of a typical thick film paste. Courtesy of Electro-Science Laboratories, Inc.

deposition and firing the ink produces a conductor pattern of metal particles surrounded by glass, the glass bonding the mixture to the substrate. Conductor materials include the precious metals silver and gold, as well as alloys of silver—palladium, gold-platinum, and gold-palladium (1–5). The conductor ink is selected on the basis of cost, compatibility with resistor and dielectric pastes, packaging concept, and individual requirements. Conductor pastes have a very high percentage of metals in comparison to glass frit and organic material. Usually only enough glass is added to the mixture to allow adequate bonding of the conductor pattern to the substrate.

An important conductor property is adhesion of the conductor to the substrate, since films receive significant stresses from the flexing of attached leads (6, 7). Peel strength is tested by pulling a ribbon lead which has been soldered to a screened and fired test pad. The constant perpendicular force that must be applied to separate the film from the substrate divided by the pad width defines peel strength, which is on the order of 100 lb/in. for palladium-silver, a material with excellent properties. A satisfactory value for thick film conductor adhesion is 25 to 35 lb/in. of width.

Conductor resistivity also must be considered when discussing conductor pastes. Purely metallic compounds generally have lower sheet resistivities than metals in an alloy form. Most applications require sheet resistivities below 0.05 $\Omega/\square$ and low contact resistance at interfaces between metals and semiconductors. Thick film pastes must exhibit chemical stability under the operating conditions of the device allowing neither interface reactions nor metal migration. Conducting films should be stable and free of phase changes when subjected to high temperatures, yet they should allow easy deposition.

To promote solderability and bondability, to insure compatibility with various resistor and dielectric film compositions, to provide high conductivity, and to be as inexpensive as possible most conductor inks are formulated from a mixture of two or more metals. Additionally, in high-frequency applications skin effect must be accounted for, usually by specifying gold or silver conductors. Solder dip techniques normally used to reduce resistivity at low frequencies are not effective at very high frequencies.

Thick film conductor materials should be capable of being as narrow, low-resistance lines. Compositions have been developed that print 2-mil-wide conductors through etched metal masks. Conventional screens will deposit conductor lines 4 mils in width. DuPont points out that in selecting a solderable conductor composition cost, line resolution, conductivity, processing conditions, solderability and bondability, adhesion, and compatibility with thick film resistor and dielectric compositions must be considered. Conductor compositions or modified versions of standard conducting pastes are available with special properties including enhanced acid or base resistance, platability by electrode or electrodeless means, cold formability, and the ability to be etched, sprayed, dipped, or brushed. Special mixtures can be formulated from glassless blends, brazable conductor compositions, conductive epoxies, and refractory metal materials.

Sheet Properties. The thick film designer usually ignores concepts such as bulk resistivity and dielectric constant and instead compares materials in terms of sheet resistance and capacitance per unit area. These properties are useful since film materials usually are screened with uniform thickness. Consider, for example, the 1000-Ω, 4-mm-wide, 4-mm-long resistor shown in Figure 3.2. Doubling the width of the resistor while maintaining the same length effectively halves the total resistance to 500 Ω. Conversely, doubling the length of the original resistor will double its resistance to 2000 Ω. By doubling both length and width simultaneously, the two effects cancel, resulting again in a 1000-Ω resistor. Any square shape of the same thickness of this particular substance will produce a 1000-Ω resistor. Thus the resistor material can be described as having a sheet resistance of 1000 Ω/□. The question then arises if a 4-mm² resistor will produce a given resistance in a circuit, why ever use a larger device or why not specify a smaller resistor and conserve space? The answer is that power dissipation is proportional to resistor area.

Since we postulate uniform printing thickness in all thick films, other sheet properties such as capacitance per unit area become a function of the material but not of the thickness of the film. Here designers use a similar concept, capacitance per unit area.

Silver Conductor Compositions. Silver is the most inexpensive conductor compound, is easily soldered to, bonds with high strength

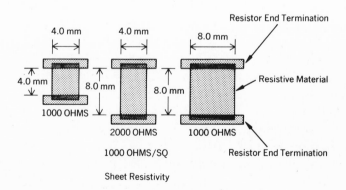

Figure 3.2. The concept of sheet resistance. Note that doubling both length and width of a thick film resistor does not change its resistance.

to the substrate, and can be fired simultaneously (cofired) with resistor compositions.

When considering skin effect, silver is seen to be a relatively excellent conductor at high frequencies. Metal migration of silver ions, the major disadvantage of using silver materials, is most noticeable in very moist environments. However, an incompatibility exists between pure silver conductor compositions and palladium-silver resistors; this can be overcome by changing from silver to silver-palladium conductor paste. The cost and high-frequency performance of silver nonetheless make it an attractive conductor material.

Characteristics of several silver pastes are tabulated in Figure 3.3.

Gold Conductor Compositions. Gold, like silver, performs excellently in high-frequency applications. Moreover, gold compositions have low sheet resistivities and outstanding bondability by die and thermocompression bonding methods and thus are widely used in chip and wired circuits. Gold can also be used in eutectic or ultrasonic bonding schemes. These conductors possess very high conductivities allowing them to be used to print fine lines where lack of resistivity is also important. However, gold film affords poor adhesion to the substrate and it is difficult to solder to gold conductors. Firing conditions must be very carefully controlled to achieve good adhesion between gold film and substrate. Moreover, since gold conductor patterns are highly soluable in conventional tin-lead solders, soldering must be done with an indium-bismith solder. Such a bond, however, is very soft and has little pull strength. In most applications, gold is wire or eutectic die bonded.

Gold pastes available with sheet resistances of 3 to 10 mΩ/$\square$ are fired at temperatures of 800 to 1000°C. Gold adheres well to alumina and steatite substrates and reasonably well to beryllia. Normal firing times for gold pastes are 10 and 20 minutes, the exact time depending on the relative proportion of gold, glass, and organics within the paste. Gold paste at $80 to $85 per troy ounce can be considered expensive.

The characteristics of gold pastes are contrasted with those of silver in Figure 3.3.

Platinum-Gold Conductor Compositions. Platinum-gold conducting compounds, extensively used in thick film applications, are

Type	Sheet Resistivity (Ω/□)	Printing and Drying	Firing Schedule (peak)	Thinner	Shelf Life	Adhesion	Bonding	Solderability
Silver (ESL-5964)	0.005	200-mesh screen. Dry at 100-125°C 15 minutes	750-850°C in air 10-15 minutes	Butyl cellosolve ESL 404	1 year	Excellent to alumina, beryllia	—	Soft solder with 3% Ag
Silver (Owens-Illinois 06103-5)	0.003 (0.3 mil thick)	220-325-mesh screen. Dry at 30°C for 2-5 minutes. Then oven dry at 100-125°C at 15 to 20 minutes.	600-900°C in air 5-15 minutes at peak temperature	OI thinner 06999	—	>2000 psi 100 mil diameter test pad	—	Good with Sn/Pb/Ag solder
Gold (ESL-8800B)	0.003 to 0.01	Dry at 120-130°C	850-1000°C in air 10-20 minutes	ESL 404 ESL 404	3-6 months	Good on alumina, fair on beryllia	Ultrasonic thermal compression	60/40 Sn/Pb, Au/Si Au/Sn
Gold (Owens-Illinois 06106)	<0.003 (0.5 mil thick)	Dry at 100-125°C 15-20 minutes	800-1000°C 5-15 minutes	OI thinner 06999	—	>2000 psi 100 mil diameter test pad	—	—

Figure 3.3. Material characteristics of representative silver and gold thick film conductor pastes.

available from most paste suppliers. These mixtures, fired from 5 to 20 minutes at temperatures of 690 to 1400°C, exhibit sheet resistivities between 16 and 120 mΩ/□. Platinum-gold films are readily thermocompression or solder bonded and have substrate adhesion characteristics similar to those of gold. Platinum-gold conductor films are compatible with silver-palladium type resistors.

Platinum-gold, expensive at $110 per troy ounce, requires precise control of firing temperature to obtain maximum substrate adhesion. Adhesion to the substrate decreases when circuits containing this material have been soldered with conventional lead-tin solders and stored for long periods at elevated temperatures. Strength can drop 50% in 1000 hours.

Palladium-Gold Conductor Compositions. Palladium-gold conductors have become very popular because they perform essentially as well as platinum-gold while delivering a considerable savings in cost. Palladium gold compounds, fired at temperatures between 750 and 1000°C, have sheet resistivities of 6 to 120 mΩ/□, are solderable using lead-tin or gold alloy solders, allow thermocompression bonding or parallel gap welding, and exhibit excellent adhesion to alumina or steatite substrates and good adhesion to beryllia. In small quantities, palladium-gold conducting inks are available at $75 per troy ounce, a savings of about 30 to 35% over platinum-gold. Care must be taken with palladium-gold systems, however, to ensure good line definition.

Palladium-Silver Conductor Compositions. Palladium-silver compositions are attractive because of their low cost. Sheet resistance of this conducting material is very high, however, on the order of 10 to 50 mΩ/□. Firing temperatures, again, are in the 700 to 1000°C range, with firing times of 10 to 20 minutes. These pastes exhibit excellent solderability with tin-lead, tin-silver, or tin-lead-silver solders. Electrical connections are made by thermocompression bonding parallel gap welding or parallel gap soldering. Palladium-silver compositions adhere excellently to alumina or steatite substrates and satisfactorily to beryllia. Although silver-palladium pastes are twice as expensive as silver, they are inexpensive compared to gold. Paste costs are about $35 per troy ounce in small quantities, dropping to $22 per troy ounce in 20-ounce lots. Since this is a silver-bearing composition, care must be taken to avoid the effects of silver migration.

Material characteristics of palladium-silver, palladium-gold, and platinum-gold are compared in Figure 3.4. Figure 3.5 is a conductor coating test pattern after being solder-dipped. This pattern is used for pull and peel adhesion tests, silver migration tests under DC polarization, printing resolution, solder leaching, and other tests.

SCREENED AND FIRED THICK FILM RESISTORS

Resistance is included in a hybrid circuit by screening and firing a high-resistivity conducting paste (8–13) or by attaching a small chip resistor. In certain circumstances using resistor chips will save the time and cost of many screening and firing cycles. The chip resistor is discussed further in Chapter 4.

Thick film resistor inks, like conductor inks, contain a suspension of metallic and glass powders within an organic vehicle. Originally carbon was used as the conducting material; its use has been discontinued. Resistor compositions now used include silver and palladium dust suspended in a glass matrix, ruthenium oxide, thallium oxide, and indium oxide. Inks of proprietary blends are also extensively marketed.

DuPont points out that the proper combination of resistor material and processing characteristics is essential to the successful manufacture of thick film hybrid microcircuitry. The resistive elements of thick film networks require the greatest precision and stability of electrical properties, although these properties often are the most sensitive to variations in processing and use. In choosing resistor compositions, fired resistor properties, design flexibility, optimum processing conditions, sensitivity to processing variables, effects of subsequent processing steps, and surface characteristics must be considered together to obtain an optimum match between resistor compositions, equipment limitations, and engineering effort consistent with circuit requirements. Commercially available resistor pastes are blended to value on an individual basis considering the screening and firing equipment to be used. Careful control must be maintained over material aspects of the screening cycle to insure that thick film inks are thoroughly mixed and have a uniformly constant

Type	Sheet Resistivity (Ω/□)	Printing and Drying	Firing Schedule (peak)	Thinner	Shelf Life	Adhesion	Bonding	Solderability
Palladium-gold (ESL 6831)	0.04–0.07 (0.8-mil film)	200–325-mesh screen. Dry at 100–125°C in air	875–1000°C 10–20 minutes	ESL 404, pine oil, BCA	6 months	Good on alumina, will bond to beryllia	Eutectic	Gold alloy solders, Sn/Pb
Palladium-gold (Owens-Illinois 06140–5)	<0.06 (0.5-mil film)	200–325-mesh screen. Level room temperature 2–5 minutes. Dry 100–125°C 15–20 minutes	750–1000°C 5–15 minutes	OI 06999	—	>2000 psi 100-mil diameter test pad	TC parallel gap welding	Sn/Pb
Palladium-silver (Cermalloy S4000)	0.04 (0.6–0.8-mil film)	200 mesh. Dry 125–150°C 20 minutes	850–975°C 10–15 minutes	—	6 months	Good on alumina and beryllia	Ultrasonic, TC wire bonding	Sn/Pb, Sn/Ag, Sn/Pb/M
Palladium-silver (Dupont DP-8430)	0.03	200 mesh	850°C	—	—	6 lb, 2.5 lb after 100 hr at 125°C	Ultrasonic aluminum, TC gold wire bonds	Sn/Pb
Platinum-gold (DuPont 7553)	0.08	—	750–1000°C	—	—	2.4 lb, 1.0 lb after 100 hr at 125°C	Ultrasonic aluminum, TC gold wire bonds	60Sn/40Pb 215°C

Figure 3.4. Material characteristics of representative alloy thick film conductor pastes.

32

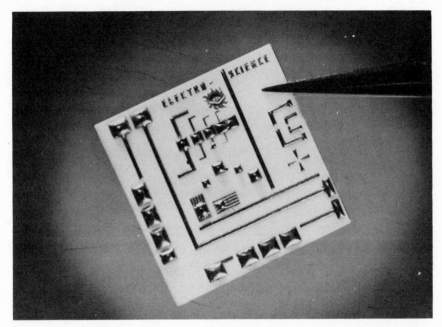

Figure 3.5. Conductor coating test pattern. Courtesy of Electro-Science Laboratories, Inc.

viscosity. This is important since the amount of paste deposited determines the final resistance of the screened and fired resistor. Screening equipment must be sturdy to insure repeatability among subsequently screened resistors.

Furnace temperature should be controlled to $\pm 2°C$ across the belt which holds the substrate, with belt speed accuracy held to $\pm 1\%$. The direction of screening by the squeegee will produce a marked effect on averge sheet resistance and variability of sheet resistance from one resistor to the next. Resistor shape will also determine final screened and fired values. Resistor paste data sheets should be consulted to determine whether a given resistor paste is compatible with the resistor end terminations. Figure 3.6 presents several screened and fired thick film resistors and conductors.

Resistivity. Overall resistor properties are determined by intensive parameters such as the bulk material resistivity, the size and shape of the resistor, and the interaction between resistive and end-termi-

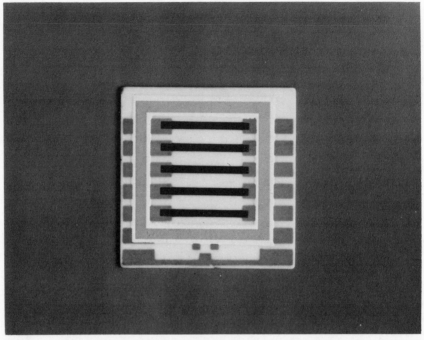

Figure 3.6. Screened and fired thick film resistors and conductors. Courtesy of Electro-Science Laboratories, Inc.

nation materials. Resistor pastes are available with sheet resistivities between 1 Ω and 10 M Ω per square. Most thick film designs use no more than two or three sheet resistivities. Resistors are commonly available with screened and fired tolerances of $\pm$ 10%, but tighter tolerances can be obtained by trimming a resistor to value after firing. Mixtures of palladium, palladium-oxide, silver, borosilicate glass frit, and an organic vehicle form a common class of thick film resistor paste. After screening, this resistor composition is baked and fired in air at 780°C. During the firing phase, the organic vehicle is driven off and palladium oxides are formed. Palladium plus silver plus the glass frit chemically combine at 780°C to produce a palladium-silver compound, palladium oxide, free palladium, and glass frit. Thus several conducting properties are responsible for conduction in the palladium thick film resistor.

For a typical thick film screened composition material, resistivity as a function of temperature follows approximately

$$\rho = TAe^{-\alpha/KT}$$

where T is temperature in °K, A is a material constant, K is Boltzman's constant, and α is the activation energy. Resistivity as a function of temperature shows a minimum at $T = \alpha/K$. Thus thick film materials can show both positive and negative temperature coefficients of resistance (TCR). Typically a high metal-to-metal ratio and high glass content within the film paste produce a positive TCR, whereas a low metal-to-oxide ratio and low glass content will produce a negative TCR.

Resistor Tolerance. Variables that affect the precision of screened and fired resistor values are substrate and screen mesh uniformity, screen emulsion and tension, physical location of resistors on the thick film substrate, and the shape and size of resistors. Tolerances also are determined by furnace firing profile and by atmosphere, contamination of the resistor paste by trapped gasses, chemical resistor paste composition, selection of material used for resistor terminations, and the composition of the substrate. Most important is uniformity of print thickness. Screened and fired resistors can be produced with initial resistance tolerances of ±20 or ±10% with three sigma confidence levels.

Temperature Coefficient of Resistance (TCR). The temperature coefficient of resistance is measured in parts per million resistance change per degree centigrade change in temperature. A typical TCR is ±100 ppm/°C. Resistor pastes usually exhibit a metal-like positive TCR for resistivities below 3.5 or 4 kΩ. Above this figure the TCR becomes negative, that is, resistance drops with increasing temperature.

The TCR of the blend can be altered by adding small amounts of oxides such as manganese oxide and nickle dioxide which have TCRs of −10,000 ppm/°C or by specially controlling the firing temperature of the film.

Noise Figure. Resistor noise is inversely proportional to the volume of resistor material used. Noise figure, standardized for a resistor test pad 0.2 × 0.1 in., is measured in units of decibels per decade with a Quantec noise analyzer.

Noise values are also a function of the composition of the resistor paste and generally increase with increasing sheet resistivity. At lower resistivities, noise values of −25 dB/decade are typical; values increase to +10 dB/decade at 100 kΩ/□. Low noise pastes of 0 dB/ decade at 100 kΩ/□ are also available.

Drift. Drift refers to a change in resistor characteristics over a time of at least 1000 hours. Occurring because of thermal expansion and chemical changes within the film, drift typically alters resistance by less than 2% after several thousand hours of dissipating 25 W/in.² at 150°C. Proprietary formulations will often drift less. DuPont, for instance, guarantees Birox® stability after 1000 hours to be better than 0.2%. For their Certi-Fired® line DuPont guarantees that drift will not exceed 1% under similar conditions. Other manufacturers such as Airco Speer, Alloys Unlimited, Bournes, Cermalloy, EMCA, and ESL guarantee similar drift characteristics.

Over a wide resistance range, increasing resistor temperature without increasing power dissipation increases resistance, whereas increasing resistor power dissipation under constant temperature conditions decreases resistance. These effects fortunately tend to mutually compensate. For resistivities below 1000 Ω/□ temperature effects are the major drift mechanism. Between 1000 and 15,000 Ω/□ the passage of current through the resistor determines long-term stability. Above 15 kΩ/□ voltage gradient effects are dominant.

Figure 3.7 is a tabulation of material characteristics of representative thick film resistor pastes.

Costs. Typical resistor paste cost figures are provided by Cermalloy. A premium paste with maximum TCR of ±50 ppm/°C costs $150 per troy ounce in small quantities and $125 in quantities of 20 ounces or more. An increase in allowable TCR to ±100 ppm drops paste prices to $125 and $100, respectively. Doubling TCR to ±200 ppm/°C reduces prices to $75 per troy ounce and $60 or $45 per troy ounce depending on resistivity, higher resistance pastes being cheaper.

Type	Sheet Resistivity (Ω/□)	TCR (ppm/°C)	Noise Quantec (dB)	Drift ΔR% 1000 hr	VCR (ppm/V/in.)	Printing and Drying	Firing Schedule (peak)	Conductor Terminations
DuPont Birox 1000 series	10–1M	<100 25–125°C	−30 at 100 Ω/□ 0 at 100KΩ/□	<0.2%	<−20 at 10K Ω/□	165- or 200-mesh screens. Level at room temperature for 5 minutes. Dry at 100–150°C 10–15 minutes	760–850°C 6–12 minutes	Pd/Ag Pd/Au Au Pt/Au
DuPont Certi-Fired 7800 series	10–100K	<300 25–125°C	−18 at 100 Ω/□ +22 at 100K Ω/□	<1%	<−50 at 10K Ω/□	Same	700–780°C 3–10 minutes	—
EMCA Firon series	30–1M	<130 −55–+200°C	−28 for 100 Ω/□ +12 at 100K Ω/□	<0.5%	—	200-mesh screen. Usual preheating and drying cycles	1800–1900°F 10–25 minutes	—
ESL 3800 series	50–10M	±100 <1000 Ω/□ ±50 1000–100,000 Ω/□	−20 to −26 for 100 Ω/□	<0.2%	60–100 in range 100K, 200K Ω/□	200-mesh screen. Level at room temperature for 5 minutes. Dry at 100–125°C 15 minutes	980–1050°C 15 minutes	Pd/Au, Pt/Au, Au

Figure 3.7. Material characteristics of representative thick film resistor pastes.

RESISTOR TRIMMING

As-fired resistors generally differ from their design values by ± 5 to $\pm 25\%$. Several trimming methods are employed to adjust resistors to more precise tolerances. These include airbrasive, laser, ultrasonic, thermal, and high-voltage pulse trimming.

Airbrasive Resistor Trimming. With the Airbrasive approach to thick film resistor trimming, a high-velocity stream of 300-mesh alundum powder is air blasted onto a thick film resistor, eroding portions of the resistor and thereby raising its resistance. During airbrasion, resistance is monitored by probes connected to a resistance bridge. When the correct value of resistance is obtained, the bridge circuitry activates a solenoid valve, stopping the airbrasive process. Cutter nozzles are available to remove resistor materials in widths of 10 to 125 mils. The amount of material removed in a given period of time is determined by gas pressure, abrasive particle flow rate, speed of the nozzle across the resistor, nozzle configuration, resistor hardness, and thickness of the resistor material. Resistors have been trimmed to values of $\pm 0.1\%$ over a range of 10Ω to $1\ M\Omega$ with this method. Production line trimmers can do 4000 trims/hour. Airbrasive trim systems including cutting nozzle, measurement bridge, substrate positioner, dust catcher, and solenoid controls are available for $3000 to $6000 (14–19).

Figures 3.8, 3.9, and 3.11 show airbrasive trimming equipment; Figure 3.10 is a photograph of a cermet resistor after trimming.

Laser Resistor Trimming. Pulsed and Q-switched lasers, such as that in Figure 3.12, are also used to trim resistors (20–22). However, because of high initial equipment cost, laser trimming generally is used only in production applications. Laser trim rates, matching those of the airbrasive technique, can be on the order of 4000 resistor trims/hour. During laser trimming, the edges of the cut are fused and sealed, thus protecting the resistor against future degradation. (The airbrasive technique leaves an exposed edge, which is susceptible to environmental contamination.) A second advantage is that the laser may be switched off much more quickly than an airbrasive stream. Thus theoretically laser trimming is capable of greater accuracy than airbrasion. Third, laser trimming is an inherently

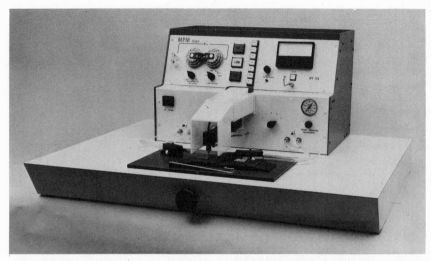

Figure 3.8. Airbrasive resistor trimmer showing cutting nozzle, substrate holder, test probes, and feedback controller. Courtesy of MPM Corporation.

clean operation, whereas the abrasive technique requires substrate cleansing to insure a dust-free thick film circuit. Fourth, the laser trim system may be used for trimming active circuits because the laser system does not generate a dust fog and electrostatic charges due to particle motion. Furthermore, since the laser is not a mechanical device, no wearing of moving parts occurs.

Laser trimming, easily computer controlled, can be done either in clean rooms or in ordinary manufacturing environments. Trim cuts that optimize resistor power dissipation capabilities, not possible with airbrasion, are easily done by laser. Sharp laser beam definition results in predictable TCR and drift coefficients. Because some lasers operate in the near infrared region, laser trimming is possible even though the hybrid circuit is glass encapsulated or overglazed. Thus the hybrid circuit can be laser trimmed during active functional testing. All available types of resistor compositions deposited on alumina can be trimmed, as can gold or cermet conductor materials. Owing to the precise nature of the laser beam, energy is applied to only a small area, avoiding damage to adjacent components and increasing hybrid circuit yield. The major disadvantage of the laser system, again, is high initial equipment cost; operating costs, how-

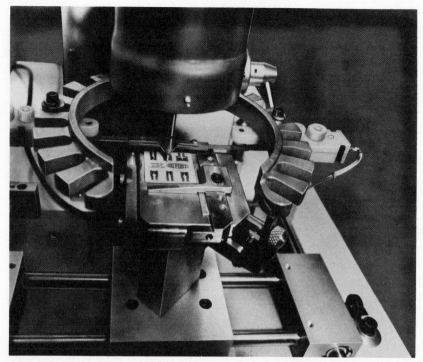

Figure 3.9. Airbrasive trimming. Courtesy of S. S. White Division of Pennwalt Corp.

ever, are low for example, a KRT YAG laser in trimming service costs about $0.25/hour for electrical power and lamps.

Ultrasonic Resistor Trimming. Resistor values also may be increased by removing resistor material with an ultrasonically vibrated 1 to 4 mil diameter diamond chisel. Its piezoelectrically activated cutting tip removes resistor material cleanly and evenly. As with airbrasive and laser methods, a monitoring device stops the trimming process when correct resistance is achieved.

Ultrasonic trimming such as that done with the chisel in Figure 3.13 is cleaner than airbrasion trimming since no dust other than that of removed resistor material is produced. This technique has lower initial cost than laser trim.

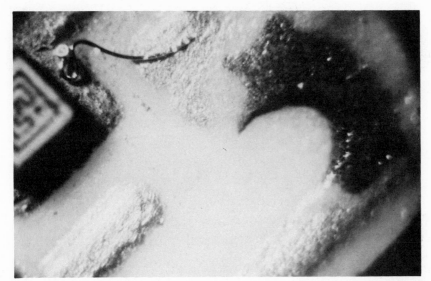

Figure 3.10. Cermet resistor after trimming. Courtesy of S. S. White Division of Pennwalt Corp.

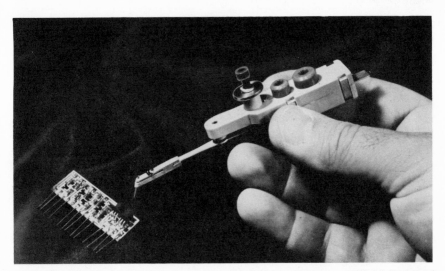

Figure 3.11. Resistor trimming monitor probe. Courtesy of S. S. White Division of Pennwalt Corp.

41

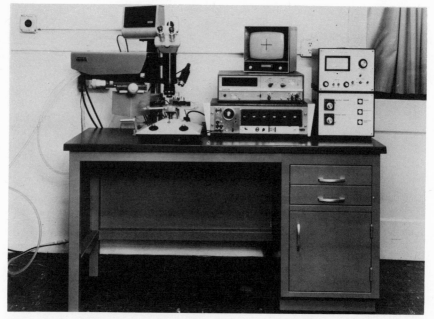

Figure 3.12.   Micromachining laser for resistor trimming. Courtesy of Korad Dept., Union Carbide Corp.

Thermal Trimming of Resistors. Thermal trimming (23) changes film resistance by permanently altering the physical properties of the thick film resistor. The resistance of an uncoated resistor can be raised by passing a large heating current through the device, oxidizing its surface. A coated resistor, however, is not readily oxidized; moreover, the adjusting current anneals the film and thereby lowers resistance. This difference is best kept in mind when resistor films are initially screened and fired, since coated resistors should be deposited at higher than design value and uncoated at lower resistance to accommodate thermal trimming. Thermal trimming action is stopped automatically by a resistance monitoring device as correct resistance is approached.

Since thermal trimming does not remove resistor material, no loss occurs in power dissipation capability. In contrast, laser and airbrasive trims usually remove 10 to 25% of the resistor material, sacrificing a like amount of power dissipation rating. Airbrasive

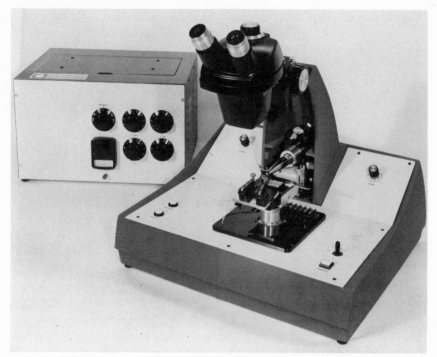

Figure 3.13. Ultrasonic trimmer and control unit. Courtesy of Axion Corp.

and laser methods also produce points of high electric field strength at the corners of resistor trims, which can be shown statistically to shorten resistor life. The thermal trim technique has been used to adjust resistors to accuracies as high as 0.01%. Depending on the tolerances required, production rates can be comparable to airbrasive systems. Thermal trimming is a nondestructive approach because it does not remove film material and is exceptionally clean since neither airbrasive grit nor resistor dust is present.

Inexpensively priced thermal trimming equipment costs less than comparable airbrasive systems. Being a stressing technique, resistor reliability is enhanced after thermal trimming. The physical properties of the resistor are of less importance since this technique alters material properties rather than the physical shape of the resistor. Thermal trim can also be used to produce resistor pairs with matched temperature coefficients of resistance. Figure 3.14 is a photograph of a thermal resistor trimmer.

Figure 3.14.   Thermal resistor trimmer. Courtesy of Data Systems Corp.

High-Voltage Pulse Trimming. Applying high-voltage, high-frequency pulses to a thick film resistor will alter the bond structure between particles of thick film material, changing the electrical resistance of the film (24). This technique has advantages similar to those of the thermal trim; it is inherently clean and nondestructive. Proponents of high-voltage pulse trimming claim a low noise figure for resistors trimmed by this method. However, the cost of required high-voltage equipment raises the cost of apparatus over that required for thermal trimming. Care also must be taken to avoid degeneration of surrounding components due to the presence of high-voltages.

DIELECTRIC FILMS

Screened and fired dielectric films (25, 26) composed of glass and ceramic frits contained within an organic vehicle commonly insulate and isolate layers of conductors in multilayer arrays, encapsulate circuit elements, and serve as capacitor dielectrics for screened and fired thick film capacitors. A pinhole-free 1.5-mil dielectric surface usually can be obtained by screening and firing two layers of insulating film. Commonly available dielectric films fire between 750 and 1000°C, but special films for crossover applications fire at 1300 to 1500°C and glazes fire at a relatively cool 500 to 540°C. Firing times here are usually about 10 to 15 minutes. A typical dielectric film has paste viscosity of 150 × 10³ cps and as-fired dielectric strength of 300 to 500 V/mil.

Figure 3.15 is a tabulation of dielectric paste characteristics; dielectrics are rated according to dielectric constant and dissipation factor. Those with the highest values in each category are called high K and high Q, respectively.

High K Dielectrics. The dielectric constant or K of thick film capacitor materials ranges between 5 and 1500. High K materials (dielectric constants 300 to 1500) made of ferroelectric powders and glass find application as dielectrics in bypass, blocking, and coupling capacitors. The actual capacity obtained with a paste of this type depends largely on processing conditions—especially firing temperature—and also on the type of electrode material used. DuPont DP8229, for instance, shows a dielectric constant of 300 when fired at 760°C which increases to 800 when fired at 1000°C. Voltage ratings of 50 to 100 and capacitance densities to 80,000 pF/in.² can be obtained with this material.

High Q Compositions. When designing high-frequency tuned circuits, a high Q characteristic is more important than high dielectric constant. Here good temperature stability and minimum dissipation is of much greater importance than capacitance density. Glass dielectrics with temperature characteristics of 1% are available with Q's between 400 and 850 at 1 MHz. These produce capacitors between 2 and 100 pF, which are used for RF, IF, and video amplifiers. Compatible with all capacitor plate materials, except perhaps palladium gold, these pastes have dielectric constants of about 10 whether measured at 1 or 1000 kHz.

Thick Film Crossover and Multilayer Compositions. Dielectrics often are deposited to provide insulation between two layers

Type	Dielectric Constant (K)	Sheet Capacitance (pF/in.²/mil)	Dielectric Strength (V/mil)	Insulation Resistance (Ω/in.²/mil)	Dissipation Factor	Temperature Coefficient	Screening Drying	Firing Schedule (peak)
Owens-Illiniois 06201-S	6	1,500	>1,000	$>10^{13}$ 10 V	<0.002 25°C 1kHz	±4% −50 to +150°C	165–200 mesh. Print one or two coats with two wet passes per coat. 165 or 200 mesh screen. Level at 30°C 5 minutes. Dry at 100–125°C 15–20 minutes	875°C 10 minutes
Owens-Illiniois 06220-S	20	5,000	>300	$>10^{10}$ 10 V	Same	±6% −50 to 150°C		
Owens-Illinois 06275-S	130	30,000	>300	$>10^{11}$ 10 V	0.008	−50 to 0°C +10% 0 to 150°C ±100 ppm/°C		875°C 10 minutes
DuPont (K500) 8229	300–800 (1 kHz)	25,000–85,000	—	$>10^{11}$ 10–100 V	1–2% 1 kHz	±20% −50 to +100°C	200 mesh. Double print. Dry 10–15 minutes at 125–150°C after each print	850–1000°C 10 minutes
ESL (K1000) 4510	1,000 ±300	70,000–175,000	—	$>10^{9}$ 50 V	2.5–4% 25°C 1 kHz	—	Same as above	925–1050°C 10–20 minutes
DuPont (K1200) DP 8289	800–1,200	52,000–160,000	—	$>10^{9}$ 100 V	<3.5% 25°C 1 kHz	±10% 0 to 105°C	Same as above	1050°C 10 minutes

Figure 3.15. Material characteristics of representative thick film dielectric pastes.

of thick film conductors. As multilayer hybrid thick film circuits become more popular, the role of the thick film crossover dielectric is becoming more important. Crossover dielectrics, fired at about 850°C, are chosen to have relatively low dielectric constants, that is, under 10, and dissipation factor less than $1\frac{1}{2}\%$. These materials must be compatible with all conductor compositions and must not soften with subsequent firings. Crossover capacities are usually less than 2 pF/crossover.

Two types of dielectrics are commonly used: glass and crystallizable glass. Glass dielectrics will resoften during subsequent refirings. Crystallizable glass dielectrics, however, do not exhibit this property, since the material partially crystallizes during firing and does not permit the "swimming" of conductors even at refirings of almost 1000°C. Highly complex multilayer structures are being constructed with these dielectrics with very high yields. Five- and six-level structures with more than 400 or 500 crossovers per square inch of substrate are attainable with substrate yields of 95% or greater.

Resistor Encapsulating Dielectrics. Thick film screened and fired resistors often are coated with an encapsulating dielectric to protect against extremes in environment and to protect against overspray in resistor trimming. Usually a quick last step in the screening and firing process, encapsulant firings are made at temperatures of 500°C for only 1 or 2 minutes at most. Even these firings may produce resistance changes of 2 or 3%, however.

Thick Film Screened and Fired Capacitors. Parallel plate capacitors can be screened and fired using sequential screening and firing of layers of metallic, then dielectric, then metallic ink again. Dielectric constants of 1200 to 1500 will produce thick film capacitors of 100,000 to 200,000 pF/in.^{2} per layer. However, for large capacitance values thick film capacitors become relatively expensive compared to external chip capacitors.

Capacitors under 500 pF screened from 1000 K dielectrics are useful in high-frequency applications. These can be produced with tolerances of $\pm 20\%$ or, under special conditions, $\pm 5\%$ with operating voltages between 100 and 250 V. Insulation resistance is 10^9 to 10^{13} Ω with dissipation factor under 1%. Temperature coefficient of capacity (TCC) is on the order of 250 ppm/°C tracking between cofired capacitors. The Q of a thick film capacitor usually stays constant to 10 MHz and then drops with increasing frequency. For small capacitors Q's greater than 500 at frequencies to 100 MHz are available.

SCREENED INDUCTORS

Thick film circuits are typically noninductive, thus in most applications requiring inductance the circuit designer avoids using a thick film inductor. An active filter, a gyrator, RC instead of LC or RL frequency tailoring, or a chip inductor is used instead. (Chip inductors are discussed in Chapter 4.) However, by screening a spiral, the circuit designer can obtain inductances of 0.2 μH, which can be useful at UHF and microwave frequencies.

SUMMARY

Thixotropic pastes for thick film applications were discussed and the properties of silver, gold, platinum-gold, palladium-gold, and palladium-silver conductors were outlined. The characteristics of screened and fired resistors, including sheet resistivity, resistor tolerance, TCR, noise figure, drift, and costs, were considered for common resistor films.

Resistor trimming techniques such as by airbrasion, laser, ultrasonics, heating, or high-voltage pulses are suggested to bring as-fired resistor tolerances of $\pm 25\%$ to better than $\pm 5\%$.

High dielectric constant films for coupling, bypass and blocking capacitors, high-Q materials for RF capacitors, and other dielectric films to facilitate conductor crossovers and to encapsulate hybrid circuits also were discussed.

REFERENCES

1. J. R. Rairden, "Thick and Thin Films for Electronic Applications—Materials and Processes Review, "*SST*, Vol. 13, No. 1, pp. 38, 39, January 1970.
2. S. B. Ruth, "Hybrids . . . Thick and Thin," *Electronic Engineer*, Vol. 28, No. 10, p. 61, October 1969.

3. D. T. DeCoursey, "Materials for Thick-Film Technology—State of the Art," *SST*, Vol. 11, No. 6, pp. 31, 32, June 1968.
4. M. F. Romano, *Hybrid Microelectronics Review*, p. 3, July 1970.
5. A. W. Postlethwaite, "Hybrid Thick Film Printed Components—Materials and Processes," *IEEE Workshop on Thick Film Hybrid IC Technology*, pp. 5-2, 5-3, 5-8, 5-9, March 1968.
6. C. J. Peckinpaugh and R. L. Tuggle, "Thick Film Adhesion—Evaluation and Improvement," *Proc. 1968 Hybrid Microelectronics Symposium*, pp. 417–423, October 1968.
7. W. A. Crossland and L. Hailes, "Thick Film Conductor Adhesion Reliability," *1970 International Hybrid Microelectronics Symposium*, pp. 3.3.1–3.3.13, November 1970.
8. G. D. Lane, "High Stability Thick-Film Resistors for Commerical Applications," *SST*, Vol. 11, No. 6, pp. 45–48, June 1968.
9. D. L. Herbst, "Composition of Thick Film Resistors," *Proc. 1968 Hybrid Microelectronics Symposium*, pp. 173–178, October 1968.
10. S. J. Stein, J. B. Garvin, and M. Vail, "Thick Film Resistor Pastes for High Performance Use," *Proc. 1969 Hybrid Microelectronics Symposium*, pp. 91–110, September 1969.
11. N. Span, "Design of Hybrid Resistors," *1970 International Hybrid Microelectronics Symposium*, pp. 7.4.1–7.4.10, November 1970.
12. T. Kubota and E. Sugata, "Variations of Electrical Characteristics with Basic Components in Pd/Ag Thick Film Resistors," *1970 International Hybrid Microelectronics Symposium*, pp. 8.6.1–8.6.16, November 1970.
13. H. Isaak, "Voltage Coefficient of Resistance of Thick Film Resistors," *1970 International Hybrid Microelectronics Symposium*, pp. 8.7.1–8.7.6, November 1970.
14. G. Thompson, III, "Air Abrasive Resistor Trimming," *SST*, Vol. 13, No. 4, p. 69, April 1970.
15. Postlethwaite, *op. cit.*, pp. 5–4, 5–9.
16. H. E. Dietsch, "Manufacturing Equipment for Large Volume Production of Hybrid IC's," *IEEE Workshop on Thick Film Hybrid IC Technology*, pp. 3-3, 3-4, 3-5, 3-8, 3-9, March 1968.
17. P. J. Sanders, "Process Information for the Design of Automatic Resistor Trimming Equipment," *Proc. 1969 Hybrid Microelectronics Symposium*, pp. 197–200, September 1969.
18. S. V. Caruso and R. V. Allen, "Development of Trimming Techniques for Microcircuit Thick-Film and Thin-Film Resistors," *1970 International Hybrid Microelectronics Symposium*, pp. 3.7.1–3.7.10, November 1970.
19. D. Ironside, "A New Low-Cost Resistor Trimming Bridge Using a Dual-Null Circuit," *1970 International Hybrid Microelectronics Symposium*, pp. 5.4.1–5.4.6, November 1970.
20. R. L. Waters and M. J. Weiner, "Resistor Trimming and Micromachining with a YAG Laser," *SST*, Vol. 13, No. 4, pp. 43–49, April 1970.
21. F. P. Burns, "Laser Trim Resistor Trimming System," *SST*, Vol. 13, No. 4, p. 70, April 1970.

22. G. B. Stone, "Programmable Continuous Trimming: A Systems Approach to High Speed Laser Trimming of Hybrid Microcircuits," *Proc. 1969 Hybrid Microelectronics Symposium*, pp. 177–183, September 1969.

23. J. F. Talbutt, Jr., "Thermal Trimming of Film Resistors," *SST*, Vol. 13, No. 4, pp. 68–69, April 1970.

24. F. J. Pakulski and T. R. Touw, "Electric Discharge Trimming of Glaze Resistors," *Proc. 1968 Hybrid Microelectronics Symposium*, pp. 145–152, October 1968.

25. Rairden, *op. cit.*, p. 37.

26. DeCoursey, *op. cit.*, p. 32.

Chapter 4

CHIP COMPONENTS

Hybrid microelectronic technology is more flexible than the mono-lithic integrated circuit art since it allows the use of discrete com-ponents individually selected from an extensive variety of chip capacitors and resistors. Possible components are silicon transistors, diodes, and monolithics and relatively small inductors and trans-formers; all are packaged in a form compatible with the thick film hybrid.

This chapter considers the electrical and physical properties of chip resistors and capacitors, the various types of dielectrics in use, and the effects of temperature, voltage, frequency, and time on capacitance. Capacitor Q and reliability as well as the properties of small inductors and transformers are mentioned. Applications of, and design alternatives for, microcircuit inductors are proposed. Means of specifying and procuring chip semiconductors, suggestions for choosing a vendor, and a brief note about test procedures are included.

The major chip package configurations—conventional chips, flip chips, beam lead devices, ceramic flip chips, and miniature packages—are described.

Thermocompression and ultrasonic bonding, parallel gap

soldering and welding, reflow soldering, back bonding schemes, laser and electron beam welding, flip chip face bonding, beam lead and ceramic flip chip attachments, and many other bonding methods are also explained; reliability, cost, and production considerations are noted throughout.

CHIP RESISTORS

Resistors are incorporated into a hybrid thick film circuit by screening and firing resistor paste or by bonding a resistor chip to the circuit substrate and then interconnecting its terminals to the rest of the hybrid (1). Although the screening and firing technique is usually simpler and less expensive, if few resistors of widely divergent values are called for in the design or if subsequent changes in resistor values may be required, then using chip resistors will eliminate multiple screening and firing steps or will eliminate the need for recalibrating the screening and firing process.

Chip resistors can be obtained for use at frequencies to the microwave range. Figures 4.1 to 4.4 show resistor chips and present electrical characteristics of chip resistors designed for thick film hybrid applications.

CHIP CAPACITORS

The screening and firing of capacitors is a multiple-step process in which a lower capacitor plate is screened on the substrate and then fired. The process is repeated to deposit a dielectric and an upper conductor. The resultant capacitor is finally encapsulated with a low firing temperature dielectric film. Despite the effort involved, the range of capacitance values possible with this technique is limited compared to those of chip capacitors. Thus chip mos, ceramic, or tantalum capacitors (Figures 4.5 to 4.10) are usually preferred over screened and fired devices (2–6).

The state-of-the-art in chip capacitors is well advanced; numerous capacities, Q's, dielectric constants, and voltage ratings are available. Capacities to several hundred microfarads obtained from small tantalum-slug devices are used in coupling and bypass applications. The ceramic chip is popular for signal processing because of

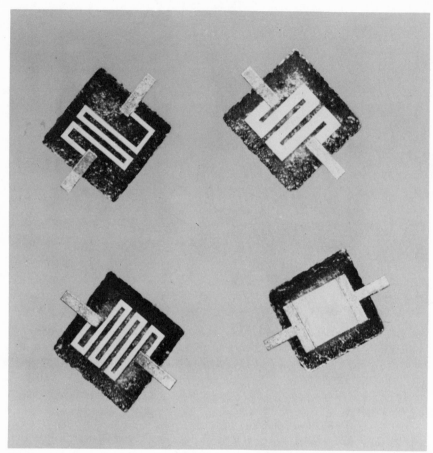

Figure 4.1. A series of 50-mil square unencapsulated thin film chip resistors for thick film hybrid applications. Courtesy of Motorola Semiconductor Products, Inc.

its ruggedness, available capacitance range to 5 mF, relatively high ratio of capacitance to unit volume, and reasonable cost.

Three basic ceramic capacitors are common: single layer thin film chips, multilayer ceramic chips, and screened-on-glass or ceramic types. Although the multilayer type is the most expensive of the group, it is the most popular because of its high capacitance per unit area. Single layer chips typically used in an opposed electrode configuration, although requiring the bonding of an extra lead and capable of only small capacities compared to multiple-layered chips,

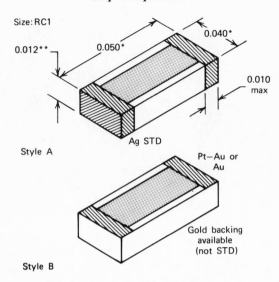

Figure 4.2. MDI Cermet chip resistor specifications. Courtesy of Monolithic Dielectrics, Inc.

Description. MDI chip resistors are microminiature devices designed for use in thick film, thin film, and other hybrid applications. A variety of terminations are available to solve the engineers' most complex bonding problems. These resistors are ideal where small size, high stability, and superior performance are required at a reasonable price.

Specifications

Electrical. Resistance range: 10 Ω to 5 MΩ; tolerance: 20%, 10% (STD), 5%, 2%, 1% available; Power rating: 100 mW; short-time overload: less than 0.50% change when tested per Mil-R-10509; Temperature coefficient of resistance: ±300 ppm/°C, ±150 ppm/°C (STD), ±50 ppm/°C available (as measured from −55 to +125°C with +25°C as reference).

Mechanical: Configuration: please refer to diagrams; construction: 96% alumina substrate with proprietary MDI cermet resistance element; termination: MDI microchip resistors are available in two standard terminations designed to be compatible with most applications. The unique Monolithic Dielectrics, Inc., wrap-around silver termination (style A) is recommended for reflow soldering. Their top surface termination (style B) is available in either gold for thermocompression bonding or platinum gold for ultrasonic bonding or strap reflow soldering. Other terminations as well as a gold backing for eutectic chip attachment are available.

Environmental. Operating temperature: −55 to +125°C; moisture resistance: less than 1% change when tested per method 106 of Mil-Std-202; life: less than 1% change when tested per method 108D (+85°C) of Mil-Std-202; noise: please refer to typical curve.

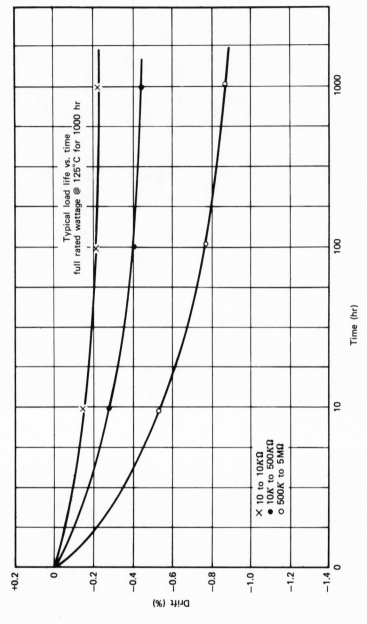

Figure 4.3. Drift characteristics—chip resistors. Courtesy of Monolithic Dielectrics, Inc.

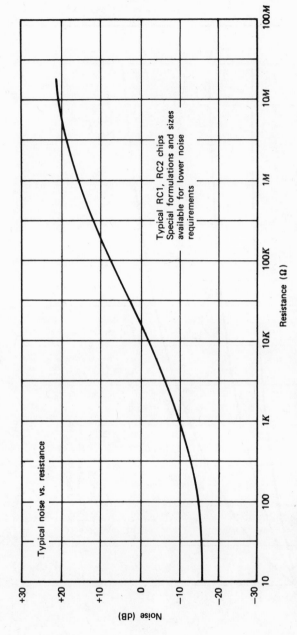

Figure 4.4.   Noise versus resistance for chip resistors. Courtesy of Monolithic Dielectrics, Inc.

56

Figure 4.5. Ceramic chip capacitors for thick film applications. Courtesy of U. S. Capacitor Corp.

retain their popularity because of very attractive prices. The single layer chip is available in a flip chip version, but the cost is higher and the range of available capacitor values smaller.

Ceramic dielectrics are divided by dielectric constant into two groups or classes: class 1 (or low K, NPO) dielectrics are larger, more expensive, and electrically of higher quality than those of class 2 (high K, K1200).

For a monolithic ceramic capacitor

$$C = \frac{AKN}{4.45t} \times 10^{-6}$$

where C = capacitance (μF)
A = the capacitor plate area (in.^{2})
K = the dielctric constant of the material

Figure 4.6. Miniature high-voltage monolithic construction capacitors. Courtesy
of Monolithic Dielectrics, Inc.

N = number of active dielectric layers
t = thickness of each dielectric layer (in.)

For the same capacitance value, a class 1 material will require
greater electrode area, since class 1 materials have lower dielectric
constants than do class 2 materials. Class 1 dielectrics are temperature
compensating and can be obtained with dielectric constants between
500 and 800. They exhibit relatively linear capacitance changes
with temperature. These materials are identifiable by an alpha-
numeric specification which shows the direction and slope of capacitance change with temperature; for example, N100 material has a

Figure 4.7. MOS capacitor chips for hybrid circuit applications. Courtesy of Dionics, Inc.

negative slope of 100 ppm/°C. As another example, the U. S. Capacitor Corporation manufacturers NPO material with a dielectric constant of about 30 and a temperature characteristic of ±0 ppm/°C.

General-purpose ceramics are a good compromise between electrical performance and size. Class 2 1200K ceramic capacitors, smaller than equivalent capacitance mica, porcelain, or NPO types, are used extensively in bypass, coupling, and blocking applications. However, since 1200K material is sensitive to voltage, temperature, and frequency changes, NPO ceramics should be specified in applications where high stability is required.

NPO ceramics, though much larger than 1200K chips, are smaller than mica capacitors for equivalent capacitance and voltage ratings. High-stability NPO chips, available with tolerances to ±1% in capacitances as small as 50 pF, are used in pulse and timing circuits and in tuning applications. NPO chips for high-frequency applications are characterized by high Q, stable capacitance with DC and AC voltage changes, high dielectric strength, and imperceptible long-term changes in value. Class 2 general-purpose ceramics have high dielectric constants, typically between 500 and 10,000.

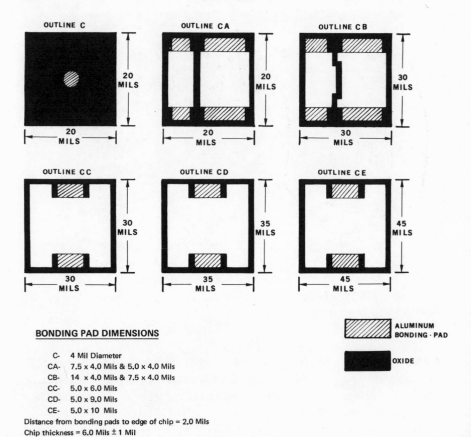

BONDING PAD DIMENSIONS

C-	4 Mil Diameter
CA-	7.5 x 4.0 Mils & 5.0 x 4.0 Mils
CB-	14 x 4.0 Mils & 7.5 x 4.0 Mils
CC-	5.0 x 6.0 Mils
CD-	5.0 x 9.0 Mils
CE-	5.0 x 10 Mils

Distance from bonding pads to edge of chip = 2.0 Mils
Chip thickness = 6.0 Mils ± 1 Mil

Figure 4.8. Physical dimensions—typical MOS single and dual unit capacitor chips. Courtesy of Dionics, Inc.

However, capacitances obtained with K1200 dielectrics do not vary linearly with temperature, change with applied AC or DC voltage level, and have higher power dissipation than do class 2 types. Nonetheless, low cost and high capacitance density make this material attractive.

Temperature Effects Class 1 and class 2 dielectrics display distinctly different characteristics over the often considered −55 to +125°C temperature range. Although all ceramic dielectrics are crystalline in nature, some crystal structures are more susceptible to change

NPO

Capacitance pF	CM-0805 0.085 × 0z055 Capacitance and Tolerance	WVDC	CM-1005 0.100 × 0z055 Capacitance and Tolerance	WVDC	CM-1007 0.100 × 0z070 Capacitance and Tolerance	WVDC	CM-1505 0.150 × 0z052 Capacitance and Tolerance	WVDC	CM-1310 0.134 × 0z102 Capacitance and Tolerance	WVDC
10	100 KM	200	100 KM	200	100 KM	200	100 KM	200		
15	150 JKM	200	150 JKM	200	150 JKM	200	150 KM	200		
22	220 JKM	200	220 JKM	200	220 JKM	200	220 KM	200	220 KM	200
33	330 JKM	200	330 JKM	200	330 JKM	200	330 KM	200	330 KM	200
47	470 JKM	200	470 JKM	200	470 JKM	200	470 JKM	200	470 KM	200
68	680 JKM	200	680 JKM	200	680 JKM	200	680 JKM	200	680 KM	200
100	101 JKM	100	101 JKM	100	101 JKM	200	101 JKM	200	101 JKM	200
120	121 JK	100	121 JK	100	121 JK	200	121 JK	200	121 JK	200
150	151 JKM	100	151 JKM	100	151 JKM	200	151 JKM	200	151 JKM	200
180	181 JK	50	181 JK	100	181 JK	100	181 JK	200	181 JK	200
220	221 JKM	50	221 JKM	100	221 JKM	100	221 JKM	100	221 JKM	200
270	271 JK	50	271 JK	50	271 JK	100	271 JK	100	271 JK	200
330	331 JKM	25	331 JKM	50	331 JKM	50	331 JKM	100	331 JKM	200
390	391 JK	25	391 JK	50	391 JK	50	391 JK	50	391 JK	100
470	471 KM	25	471 JKM	25	471 JKM	50	471 JKM	50	471 JKM	100
560			561 K	25	561 JK	25	561 JK	50	561 JK	100
680					681 JKM	25	681 JKM	25	681 JKM	100
820					821 K	25	821 JK	25	821 JK	50
1,000							102 KM	25	102 JKM	50
1,200									122 JK	50
1,500									152 JKM	25
1,800									182 K	25

Capacitance pF	CM-1316 0.135 × 0.165 Capacitance and Tolerance	WVDC	CM-1712 0.175 × 0.125 Capacitance and Tolerance	WVDC	CM-1922 0.193 × 0.224 Capacitance and Tolerance	WVDC	CM-2225 0.220 × 0.250 Capacitance and Tolerance	WVDC	CM-3816 0.380 × 0.157 Capacitance and Tolerance	WVDC
47	470 KM	200	470 KM	200						
68	680 KM	200	680 KM	200						
100	101 JKM	200	101 JKM	200	101 KM	200	101 KM	200		
150	151 JKM	200	151 JKM	200	151 KM	200	151 KM	200	151 KM	200
220	221 JKM	200	221 JKM	200	221 KM	200	221 KM	200	221 KM	200
330	331 JKM	200	331 JKM	200	331 JKM	200	331 JKM	200	331 JKM	200
470	471 JKM	200	471 JKM	200	471 JKM	200	471 JKM	200	471 JKM	200
560	561 JK	200	561 JK	200	561 JK	200	561 JK	200	561 JK	200
680	681 JKM	100	681 JKM	100	681 JKM	200	681 JKM	200	681 JKM	200
820	821 JK	100	821 JK	100	821 JK	200	821 JK	200	821 JK	200
1,000	102 JKM	100	102 JKM	100	102 JKM	200	102 JKM	200	102 JKM	200
1,200	122 JK	100	122 JK	100	122 JK	200	122 JK	200	122 JK	200
1,500	152 JKM	50	152 JKM	50	152 JKM	100	152 JKM	200	152 JKM	200
1,800	182 JK	50	182 JK	50	182 JK	100	182 JK	100	182 JK	100
2,200	222 JKM	50	222 JKM	50	222 JKM	100	222 JKM	100	222 JKM	100
2,700	272 JK	50	272 JK	50	272 JK	100	272 JK	100	272 JK	100
3,300	332 JKM	25	332 JKM	25	332 JKM	50	332 JKM	100	332 JKM	100
3,900	392 K	25	392 K	25	392 JK	50	392 JK	50	392 JK	50
4,700					472 JKM	50	472 JKM	50	472 JKM	50
5,600					562 JK	25	562 JK	50	562 JK	50
6,800					682 JKM	25	682 JKM	25	682 JKM	25
8,200					822 K	25	822 K	25	822 K	25
10,000							103 M	25	103 M	25

J = ±5%,   K = ±10%,   L = ±20%

Figure 4.9. Component values—thick film application capacitors. Courtesy of American Lava Corp., Subsidiary 3M Company.

K1300

Capacitance pF	CM-0804 0.085 × 0.055		CM-1005 0.100 × 0.055		CM-1007 0.100 × 0.070		CM-1505 0.150 × 0.052		CM-1310 0.134 × 0.102	
	Capacitance and Tolerance	WVDC	Capacitance and Tolerance	WVDC	Capacitance and Tolerance	WVDC	Capacitance and Tolerance	WVDC	Capacitance and Tolerance	WVDC
100	101 KM	200								
150	151 KM	200	151 KM	200						
220	221 KM	200	221 KM	200	221 KM	200	221 KM	200		
330	331 KM	200	331 KM	200	331 KM	200	331 KM	200		
470	471 JKM	200	471 JKM	200	471 JKM	200	471 JKM	200	471 KM	200
680	681 JKM	200	681 JKM	200	681 JKM	200	681 JKM	200	681 KM	200
1,000	102 JKM	200	102 JKM	200	102 JKM	200	102 JKM	200	102 JKM	200
1,500	152 JKM	200	152 JKM	200	152 JKM	200	152 JKM	200	152 JKM	200
2,200	222 JKM	100	222 JKM	200	222 JKM	200	222 JKM	200	222 JKM	200
3,300	332 JKM	100	332 JKM	100	332 JKM	200	332 JKM	200	332 JKM	200
3,900	392 JK	50	392 JK	100	392 JK	100	392 JK	200	392 JK	200
4,700	472 JKM	50	472 JKM	100	472 JKM	100	472 JKM	100	472 JKM	200
5,600	562 JK	50	562 JK	50	562 JK	100	562 JK	100	562 JK	200
6,800	682 JKM	50	682 JKM	50	682 JKM	100	682 JKM	100	682 JKM	200
8,200	822 JK	25	822 JK	50	822 JK	50	822 JK	100	822 JK	100
10,000	103 KM	25	103 JKM	25	103 JKM	50	103 JKM	50	103 JKM	100
mF										
.012			123 K	25	123 JK	50	123 JK	50	123 JK	100
.015					153 JKM	25	153 JKM	50	153 JKM	100
.018					183 K	25	183 JK	25	183 JK	50
.022							223 KM	25	223 JKM	50
.027									273 JK	50
.033									333 JKM	25
.039									393 K	25

Capacitance pF	CM-1316 0.135 × 0.165		CM-1712 0.175 × 0.125		CM-1972 0.193 × 0.224		CM-2225 0.220 × 0.250		CM-3816 0.380 × 0.157	
	Capacitance and Tolerance	WVDC	Capacitance and Tolerance	WVDC	Capacitance and Tolerance	WVDC	Capacitance and Tolerance	WVDC	Capacitance and Tolerance	WVDC
1,000	102 JKM	200	102 JKM	200						
1,500	152 JKM	200	152 JKM	200						
2,200	222 JKM	200	222 JKM	200	22 KM	200	222 KM	200	222 KM	200
3,300	332 JKM	200	332 JKM	200	332 KM	200	332 KM	200	332 KM	200
3,900	392 JK	200	392 JK	200	392 JK	200	392 JK	200	392 JK	200
4,700	472 JKM	200	472 JKM	200	472 JKM	200	472 JKM	200	472 JKM	200
5,600	562 JK	200	562 JK	200	562 JK	200	562 JK	200	562 JK	200
6,800	682 JKM	200	682 JKM	200	682 JKM	200	682 JKM	200	682 JKM	200
8,200	822 JK	200	822 JK	200	822 JK	200	822 JK	200	822 JK	200
10,000	103 JKM	200	103 JKM	200	103 JKM	200	103 JKM	200	103 JKM	200
mF										
.012	123 JK	200	123 JK	200	123 JK	200	123 JK	200	123 JK	200
.015	153 JKM	200	153 JKM	200	153 JKM	200	153 JKM	200	153 JKM	200
.018	183 JK	100	183 JK	100	183 JK	200	183 JK	200	183 JK	200
.022	223 JKM	100	223 JKM	100	223 JKM	200	223 JKM	200	223 JKM	200
.027	273 JK	100	273 JK	100	273 JK	200	273 JK	200	273 JK	200
.033	333 JKM	100	333 JKM	100	333 JKM	200	333 JKM	200	333 JKM	200
.039	393 JK	50	393 JK	50	393 JK	100	393 JK	200	393 JK	200
.047	473 JKM	50	473 JKM	50	473 JKM	100	473 JKM	100	473 JKM	100
.056	563 JK	50	563 JK	50	563 JK	100	563 JK	100	563 JK	100
.068	683 JKM	25	683 JKM	25	683 JKM	100	683 JKM	100	683 JKM	100
.082	823 K	25	823 K	25	823 JK	50	823 JK	100	823 JK	100
.10					104 JKM	50	104 JKM	50	104 JKM	50
.12					124 JK	25	124 JK	50	124 JK	50
.15					154 KM	25	154 KM	25	154 JKM	50
.18							184 K	25	184 JK	25
.22									224 M	25

Figure 4.9. *Continued.*

FEATURES

- COMPATIBILITY WITH MICROELECTRONIC CIRCUITRY.
- RELIABILITY IN A TEMPERATURE RANGE OF −55°C TO +125°C.

RATINGS

CAPACITANCE RANGE: .001 μf — 220 μf
(Extended Capacitance and Voltage Ranges available upon request.)
DC RATED VOLTAGE: 2 thru 50V
LEAKAGE CURRENT: Need not be less than 1.0μA, or .02μA x μfd x working voltage.
DISSIPATION FACTOR: Values generally do not exceed 10% except in certain low voltage devices.
OPERATING TEMPERATURE RANGE: −55°C to +125°C
DC VOLTAGE DERATING: Above 85°C. Voltage is derated linearly to 125°C.
The 125°C voltage is ⅔ of the 85°C value.

Figure 4.10.   Tantalum capacitor specifications. Courtesy of Dickson Electronics Corp.

AXIAL=X CYLINDRICAL POLAR RADIAL=Y

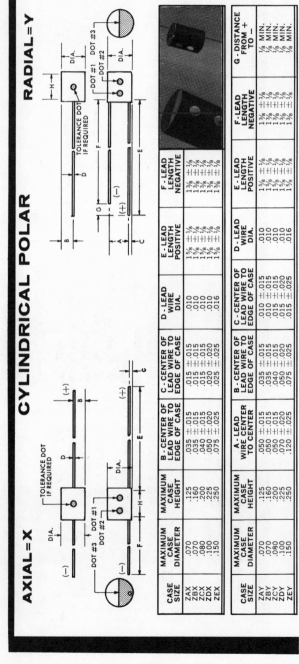

CASE SIZE	MAXIMUM CASE DIAMETER	MAXIMUM CASE HEIGHT	B - CENTER OF LEAD WIRE TO CENTER	C - CENTER OF LEAD WIRE TO EDGE OF CASE	D - LEAD WIRE DIA.	E - LEAD LENGTH POSITIVE	F - LEAD LENGTH NEGATIVE
ZAX	.070	.125	$.035 \pm .015$	$.015 \pm .015$	.010	$1\frac{5}{8} \pm \frac{1}{8}$	$1\frac{3}{8} \pm \frac{1}{8}$
ZBX	.070	.160	$.035 \pm .015$	$.015 \pm .015$	.010	$1\frac{5}{8} \pm \frac{1}{8}$	$1\frac{3}{8} \pm \frac{1}{8}$
ZCX	.080	.200	$.040 \pm .015$	$.015 \pm .015$	.010	$1\frac{5}{8} \pm \frac{1}{8}$	$1\frac{3}{8} \pm \frac{1}{8}$
ZDX	.100	.225	$.050 \pm .020$	$.020 \pm .020$	.010	$1\frac{5}{8} \pm \frac{1}{8}$	$1\frac{3}{8} \pm \frac{1}{8}$
ZEX	.150	.250	$.075 \pm .025$	$.025 \pm .025$	.016	$1\frac{5}{8} \pm \frac{1}{8}$	$1\frac{3}{8} \pm \frac{1}{8}$

CASE SIZE	MAXIMUM CASE DIAMETER	MAXIMUM CASE HEIGHT	A - LEAD WIRE CENTER TO CENTER	B - CENTER OF LEAD WIRE TO EDGE OF CASE	C - CENTER OF LEAD WIRE TO EDGE OF CASE	D - LEAD WIRE DIA.	E - LEAD LENGTH POSITIVE	F - LEAD LENGTH NEGATIVE	G - DISTANCE FROM + TO -
ZAY	.070	.125	$.050 \pm .015$	$.035 \pm .015$	$.010 \pm .015$	.010	$1\frac{5}{8} \pm \frac{1}{8}$	$1\frac{3}{8} \pm \frac{1}{8}$	$\frac{1}{8}$ MIN.
ZBY	.070	.160	$.050 \pm .015$	$.035 \pm .015$	$.010 \pm .015$	.010	$1\frac{5}{8} \pm \frac{1}{8}$	$1\frac{3}{8} \pm \frac{1}{8}$	$\frac{1}{8}$ MIN.
ZCY	.080	.200	$.050 \pm .015$	$.040 \pm .015$	$.015 \pm .015$	.010	$1\frac{5}{8} \pm \frac{1}{8}$	$1\frac{3}{8} \pm \frac{1}{8}$	$\frac{1}{8}$ MIN.
ZDY	.100	.225	$.070 \pm .020$	$.050 \pm .020$	$.015 \pm .020$	.010	$1\frac{5}{8} \pm \frac{1}{8}$	$1\frac{3}{8} \pm \frac{1}{8}$	$\frac{1}{8}$ MIN.
ZEY	.150	.250	$.120 \pm .025$	$.075 \pm .025$	$.015 \pm .025$	.016	$1\frac{5}{8} \pm \frac{1}{8}$	$1\frac{3}{8} \pm \frac{1}{8}$	$\frac{1}{8}$ MIN.

Figure 4.10. Continued

CAPACITANCE μfd	D. C. WORKING VOLTAGE							
	2 VOLTS	4 VOLTS	6 VOLTS	10 VOLTS	15 VOLTS	20 VOLTS	35 VOLTS	50 VOLTS
.001	DR001ZA* 2M	DR001ZA* 4M	DR001ZA* 6M	DR001ZA* 10M	DR001ZA* 15M	DR001ZA* 20M	DR001ZA* 35M	DR001ZB* 50M
.0015	DR0015ZA* 2M	DR0015ZA* 4M	DR0015ZA* 6M	DR0015ZA* 10M	DR0015ZA* 15M	DR0015ZA* 20M	DR0015ZA* 35M	DR0015ZB* 50M
.0022	DR0022ZA* 2M	DR0022ZA* 4M	DR0022ZA* 6M	DR0022ZA* 10M	DR0022ZA* 15M	DR0022ZA* 20M	DR0022ZA* 35M	DR0022ZB* 50M
.0033	DR0033ZA* 2M	DR0033ZA* 4M	DR0033ZA* 6M	DR0033ZF* 10M	DR0033ZA* 15M	DR0033ZA* 20M	DR0033ZA* 35M	DR0033ZB* 50M
.0047	DR0047ZA* 2M	DR0047ZA* 4M	DR0047ZA* 6M	DR0047ZA* 10M	DR0047ZA* 15M	DR0047ZA* 20M	DR0047ZA* 35M	DR0047ZB* 50M
.0068	DR0068ZA* 2M	DR0068ZA* 4M	DR0068ZA* 6M	DR0068ZA* 10M	DR0068ZA* 15M	DR0068ZA* 20M	DR0068ZA* 35M	DR0068ZB* 50M
.01	DR01ZA* 2M	DR01ZA* 4M	DR01ZA* 6M	DR01ZA* 10M	DR01ZA* 15M	DR01ZA* 20M	DR01A* 35M	DR01B* 50M
.015	DR015ZA* 2M	DR015ZA* 4M	DR015ZA* 6M	DR015ZA* 10M	DR015ZA* 15M	DR015ZA* 20M	DR015ZB* 35M	
.022	DR022ZA* 2M	DR022ZA* 4M	DR022ZA* 6M	DR022ZA* 10M	DR022ZA* 15M	DR022ZA* 20M	DR022ZB* 35M	
.033	DR033ZA* 2M	DR033ZA* 4M	DR033ZA* 6M	DR033ZA* 10M	DR033ZA* 15M	DR033ZA* 20M	DR033ZB* 35M	
.047	DR047ZA* 2M	DR047ZA* 4M	DR047ZA* 6M	DR047ZA* 10M	DR047ZA* 15M	DR047ZA* 20M	DR047ZB* 35M	
.068	DR068ZA* 2M	DR068ZA* 4M	DR068ZA* 6M	DR068ZA* 10M	DR068ZA* 15M	DR068ZA* 20M	DR068ZB* 35M	
.10	DR10ZA* 2M	DR10ZA* 4M	DR10ZA* 6M	DR10ZA* 10M	DR10ZA* 15M	DR10ZA* 20M	DR10ZB* 35M	
.15	DR15ZA* 2M	DR15ZA* 4M	DR15ZA* 6M	DR15ZA* 10M	DR15ZA* 15M	DR15ZA* 20M	DR15ZB* 35M	
.22	DR22ZA* 2M	DR22ZA* 4M	DR22ZA* 6M	DR22ZA* 10M	DR22ZA* 15M	DR22ZA* 20M	DR22ZB* 35M	
.33	DR33ZA* 2M	DR33ZA* 4M	DR33ZA* 6M	DR33ZA* 10M	DR33ZA* 15M	DR33ZB* 20M	DR33ZC* 35M	
.47	DR47ZA* 2M	DR47ZA* 4M	DR47ZA* 6M	DR47ZA* 10M	DR47ZB* 15M	DR47ZB* 20M	DR47ZC* 35M	
.68	DR68ZA* 2M	DR68ZA* 4M	DR68ZA* 6M	DR68ZB* 10M	DR68ZB* 15M	DR68ZC* 20M	DR68ZD* 35M	
1.0	D1R0ZA* 2M	D1R0ZA* 4M	D1R0ZA* 6M	D1R0ZB* 10M	D1R0ZC* 15M	D1R0ZC* 20M	D1R0ZD* 35M	
1.5	D1R5ZA* 2M	D1R5ZA* 4M	D1R5ZA* 6M	D1R5ZB* 10M	D1R5ZC* 15M	D1R5ZD* 20M	D1R5ZE* 35M	
2.2	D2R2ZA* 2M	D2R2ZA* 4M	D2R2ZB* 6M	D2R2ZC* 10M	D2R2ZC* 15M	D2R2ZD* 20M	D2R2ZE* 35M	
3.3	D3R3ZB* 2M	D3R3ZB* 4M	D3R3ZB* 6M	D3R3ZC* 10M	D3R3ZD* 15M	D3R3ZE* 20M		
4.7	D4R7ZB* 2M	D4R7ZC* 4M	D4R7ZC* 6M	D4R7ZD* 10M	D4R7ZE* 15M	D4R7ZE* 20M		
6.8	D6R8ZC* 2M	D6R8ZD* 4M	D6R8ZD* 6M	D6R8ZD* 10M	D6R8ZE* 15M	D6R8ZE* 20M		
10.	D10ZC* 2M	D10ZD* 4M	D10ZE* 6M	D10ZE* 10M	D10ZE* 15M			
15.	D15ZD* 2M	D15ZD* 4M	D15ZE* 6M	D15ZE* 10M				
22.	D22ZD* 2M	D22ZE* 4M	D22ZE* 6M					
33.	D322E* 2M	D322E* 4M						
47.	D47ZE* 2M							

NOTE: *WHEN ORDERING, INDICATE EITHER AXIAL ("X") OR RADIAL ("Y") LEAD CONFIGURATION BY PLACING "X" OR "Y" IN DEVICE CODE NUMBER SPACE NOW OCCUPIED BY ASTERISK. EXAMPLE: D2R2ZAX2M INDICATES YOU REQUIRE THIS DEVICE WITH AN AXIAL LEAD.

Figure 4.10. Continued

MECHANICAL CHARACTERISTICS

DIMENSIONS: See outline.
POLARITY: Positive lead is indicated by position of third dot and longer lead.
TOLERANCE: A color dot designating tolerance will be placed on the package if required.
LEADS: Solder-coated grade A nickel.

CAPACITANCE COLOR CODE DESIGNATION		
COLOR	1st and 2nd DOTS (NOTE 1)	3rd DOT (NOTE 2)
BLACK	0	1
BROWN	1	10
RED	2	100
ORANGE	3	1,000
YELLOW	4	10,000
GREEN	5	100,000
BLUE	6	1,000,000
VIOLET	7	
GRAY	8	
WHITE	9	

TOLERANCE COLOR CODE	
TOLERANCE	COLOR DOT
±5%	GOLD
±10%	SILVER
±20%	NONE
+40% / −20%	NONE

NOTE 1: The first and second dots designate the first and second significant figures of capacitance (in pf) respectively.

NOTE 2: The color of the third dot designates the multiplier of the significant figures and positive lead.

DOT #3

DOT #2 DOT #1

TOLERANCE

Figure 4.10. Continued

66

AXIAL=X   RECTANGULAR POLAR   RADIAL=Y

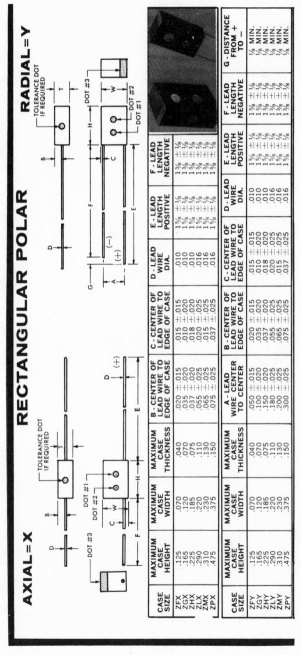

CASE SIZE	MAXIMUM CASE HEIGHT	MAXIMUM CASE WIDTH	MAXIMUM CASE THICKNESS	B - CENTER OF LEAD WIRE TO EDGE OF CASE	C - CENTER OF LEAD WIRE TO EDGE OF CASE	D - LEAD WIRE DIA.	E - LEAD LENGTH POSITIVE	F - LEAD LENGTH NEGATIVE
ZFX	.125	.070	.040	.020 ± .015	.015 ± .015	.010	1 5/8 ± 1/8	1 3/8 ± 1/8
ZGX	.165	.120	.070	.035 ± .020	.010 ± .020	.010	1 5/8 ± 1/8	1 3/8 ± 1/8
ZHX	.225	.185	.075	.037 ± .020	.018 ± .020	.010	1 5/8 ± 1/8	1 3/8 ± 1/8
ZLX	.290	.220	.110	.055 ± .025	.020 ± .025	.016	1 5/8 ± 1/8	1 3/8 ± 1/8
ZMX	.310	.230	.130	.065 ± .025	.015 ± .025	.016	1 5/8 ± 1/8	1 3/8 ± 1/8
ZPX	.475	.375	.150	.075 ± .025	.037 ± .025	.016	1 5/8 ± 1/8	1 3/8 ± 1/8

CASE SIZE	MAXIMUM CASE HEIGHT	MAXIMUM CASE WIDTH	MAXIMUM CASE THICKNESS	A - LEAD WIRE CENTER TO CENTER	B - CENTER OF LEAD WIRE TO EDGE OF CASE	C - CENTER OF LEAD WIRE TO EDGE OF CASE	D - LEAD WIRE DIA.	E - LEAD LENGTH POSITIVE	F - LEAD LENGTH NEGATIVE	G - DISTANCE FROM + TO −
ZFY	.125	.070	.040	.050 ± .015	.020 ± .015	.015 ± .015	.010	1 5/8 ± 1/8	1 3/8 ± 1/8	1/8 MIN.
ZGY	.165	.120	.070	.150 ± .020	.035 ± .020	.010 ± .020	.010	1 5/8 ± 1/8	1 3/8 ± 1/8	1/8 MIN.
ZHY	.225	.185	.075	.180 ± .025	.037 ± .020	.018 ± .025	.016	1 5/8 ± 1/8	1 3/8 ± 1/8	1/8 MIN.
ZLY	.290	.220	.110	.180 ± .025	.055 ± .025	.020 ± .025	.016	1 5/8 ± 1/8	1 3/8 ± 1/8	1/8 MIN.
ZMY	.290	.230	.130	.200 ± .025	.065 ± .025	.015 ± .025	.016	1 5/8 ± 1/8	1 3/8 ± 1/8	1/8 MIN.
ZPY	.475	.375	.150	.300 ± .025	.075 ± .025	.037 ± .025	.016	1 5/8 ± 1/8	1 3/8 ± 1/8	1/8 MIN.

Figure 4.10. Continued

CAPACITANCE	D. C. WORKING VOLTAGE							
μfd	2 VOLTS	4 VOLTS	6 VOLTS	10 VOLTS	15 VOLTS	20 VOLTS	35 VOLTS	50 VOLTS
.001	DR0012F* 2M	DR0012F* 4M	DR0012F* 6M	DR0012F* 10M	DR0012F* 15M	DR0012F* 20M	DR0012F* 35M	DR0012ZG* 50M
.0015	DR0015ZF* 2M	DR0015ZF* 4M	DR0015ZF* 6M	DR0015ZF* 10M	DR0015ZF* 15M	DR0015ZF* 20M	DR0015ZF* 35M	DR0015ZG* 50M
.0022	DR0022ZF* 2M	DR0022ZF* 4M	DR0022ZF* 6M	DR0022ZF* 10M	DR0022ZF* 15M	DR0022ZF* 20M	DR0022ZF* 35M	DR0022ZG* 50M
.0033	DR0033ZF* 2M	DR0033ZF* 4M	DR0033ZF* 6M	DR0033ZF* 10M	DR0033ZF* 15M	DR0033ZF* 20M	DR0033ZF* 35M	DR0033ZG* 50M
.0047	DR0047ZF* 2M	DR0047ZF* 4M	DR0047ZF* 6M	DR0047ZF* 10M	DR0047ZF* 15M	DR0047ZF* 20M	DR0047ZF* 35M	DR0047ZG* 50M
.0068	DR0068ZF* 2M	DR0068ZF* 4M	DR0068ZF* 6M	DR0068ZF* 10M	DR0068ZF* 15M	DR0068ZF* 20M	DR0068ZF* 35M	DR0068ZG* 50M
.01	DR012F* 2M	DR012F* 4M	DR012F* 6M	DR012F* 10M	DR012F* 15M	DR012F* 20M	DR012F* 35M	DR0102ZG* 50M
.015	DR015ZF* 2M	DR015ZF* 4M	DR015ZF* 6M	DR015ZF* 10M	DR015ZF* 15M	DR015ZF* 20M	DR015ZG* 35M	
.022	DR022ZF* 2M	DR022ZF* 4M	DR022ZF* 6M	DR022ZF* 10M	DR022ZF* 15M	DR022ZF* 20M	DR022ZG* 35M	
.033	DR033ZF* 2M	DR033ZF* 4M	DR033ZF* 6M	DR033ZF* 10M	DR033ZF* 15M	DR033ZF* 20M	DR033ZG* 35M	
.047	DR047ZF* 2M	DR047ZF* 4M	DR047ZF* 6M	DR047ZF* 10M	DR047ZF* 15M	DR047ZF* 20M	DR047ZG* 35M	
.068	DR068ZF* 2M	DR068ZF* 4M	DR068ZF* 6M	DR068ZF* 10M	DR068ZF* 15M	DR068ZF* 20M	DR068ZG* 35M	
.10	DR102F* 2M	DR102F* 4M	DR102F* 6M	DR102F* 10M	DR102F* 15M	DR102F* 20M	DR102G* 35M	
.15	DR152F* 2M	DR152F* 4M	DR152F* 6M	DR152F* 10M	DR152F* 15M	DR152F* 20M	DR152G* 35M	
.22	DR222F* 2M	DR222F* 4M	DR222F* 6M	DR222F* 10M	DR222F* 15M	DR222F* 20M	DR222G* 35M	
.33	DR332F* 2M	DR332F* 4M	DR332F* 6M	DR332F* 10M	DR332F* 15M	DR332G* 20M	DR332G* 35M	
.47	DR472F* 2M	DR472F* 4M	DR472F* 6M	DR472F* 10M	DR472G* 15M	DR472F* 20M	DR472G* 35M	
.68	DR682F* 2M	DR682F* 4M	DR682F* 6M	DR682G* 10M	DR682G* 15M	DR682G* 20M	DR682H* 35M	
1.0	D1R0ZF* 2M	D1R0ZF* 4M	D1R0ZG* 6M	D1R0ZG* 10M	D1R0ZG* 15M	D1R0ZG* 20M	D1R0ZH* 35M	
1.5	D1R5ZF* 2M	D1R5ZG* 4M	D1R5ZG* 6M	D1R5ZG* 10M	D1R5ZG* 15M	D1R5ZH* 20M	D1R5ZH* 35M	
2.2	D2R2ZG* 2M	D2R2ZG* 4M	D2R2ZG* 6M	D2R2ZG* 10M	D2R2ZH* 15M	D2R2ZH* 20M	D2R2ZL* 35M	
3.3	D3R3ZG* 2M	D3R3ZG* 4M	D3R3ZG* 6M	D3R3ZH* 10M	D3R3ZH* 15M	D3R3ZH* 20M	D3R3ZL* 35M	
4.7	D4R7ZG* 2M	D4R7ZG* 4M	D4R7ZG* 6M	D4R7ZH* 10M	D4R7ZH* 15M	D4R7ZH* 20M	D4R7ZL* 35M	
6.8	D6R8ZG* 2M	D6R8ZG* 4M	D6R8ZG* 6M	D6R8ZH* 10M	D6R8ZH* 15M	D6R8ZH* 20M	D6R8ZM* 35M	
10.	D102G* 2M	D102H* 4M	D102H* 6M	D102H* 10M	D102L* 15M	D102L* 20M	D102P* 35M	
15.	D152H* 2M	D152H* 4M	D152H* 6M	D152L* 10M	D152L* 15M	D152M* 20M	D152P* 35M	
22.	D222H* 2M	D222L* 4M	D222L* 6M	D222L* 10M	D222M* 15M	D222P* 20M	D222P* 35M	
33.	D332L* 2M	D332L* 4M	D332L* 6M	D332M* 10M	D332P* 15M	D332P* 20M		
47.	D472L* 2M	D472L* 4M	D472M* 6M	D472P* 10M	D472P* 15M	D472P* 20M		
68.	D682M* 2M	D682M* 4M	D682P* 6M	D682P* 10M	D682P* 15M			
82.	D822M* 2M	D822P* 4M	D822P* 6M	D822P* 10M				
100.	D100ZP* 2M	D100ZP* 4M	D100ZP* 6M	D100ZP* 10M				
150.	D150ZP* 2M	D150ZP* 4M	D150ZP* 6M					
220.	D220ZP* 2M							

Figure 4.10.   Continued

68

with temperature than are others. NPO materials are temperature stable since they maintain the same crystalline structure. Thus low *K*, NPO, or class 1 materials produce constant capacitances over the entire temperature range. Conversely, class 2, or K1200 material, shows a positive capacitance change versus temperature (Figure 4.11) from −55 to about +5°C, whereupon the capacitance will drop only to increase again in the vicinity of the Curie point of the material at 115°C. Here the crystalline structure changes from tetragonal to cubic. Beyond the Curie point, capacitance again begins to drop sharply. If the Curie point of a class 2 ceramic material is exceeded, there is a net increase in capacitance of about 5%, even after cooling, since the cubic structure of the ceramic does not immediately relax back to tetragonal. This relaxation comes with time: capacitance drops at the rate of 1 to 2%/hour/decade (i.e., after 1 hour, the drop is 1 or 2%, 10 hours later it is another 1 or 2%, and so on). The aging process occurs every time the dielectric is heated above the Curie point. These aging effects do not occur in class 1 materials.

Class 2 materials of barium titanate have room temperature dielectric constants of 1500 to 3000, which remain stable to ±100°C. This dielectric constant increases to 10,000 at 120°C but falls rapidly beyond 125°C. Below the Curie point Perovskite materials such as barium titanate are ferroelectric, whereas above they are paraelectric.

Dissipation Factor and Insulation Resistance. Schematically, capacitor dissipation factor or *Q* can be represented as resistance in series with the capacitance. Insulation resistance can be depicted as resistance in parallel with the capacitance (Figure 4.12). For capacities above 1000 pF for K1200 material an implicit relationship between insulation resistance and capacitance, megohms × micro-

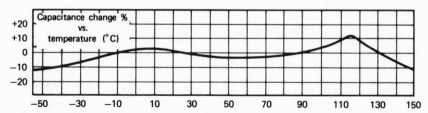

Figure 4.11. Capacitance change versus temperature, K1200 material. Courtesy of Monolithic Dielectrics, Inc.

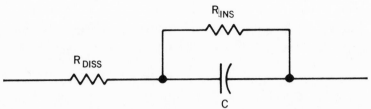

Figure 4.12.   Schematic representation of capacitor dissipation and insulation
resistance.

farads = 300, holds quite well. Below 1000 pF the insulation resistance
of the encapsulant (typically on the order of 1,000,000 MΩ) is the
dominant charge leakage path. With 1200 material insulation
resistance decreases with increasing temperature. (Typically, insulation resistance will drop 90% with 100°C increase).

The insulation resistance of NPO material decreases with
increasing temperature. Dissipation factor decreases with increasing
temperatures for both class 1 and class 2 materials. For K1200
dielectric dissipation factor will be 1% at 25°C, dropping to 0.2%
at 125°C. NPO material has a dissipation factor of 0.05% at 25°C,
with a small drop as temperature is increased to 125°C. Essentially,
however, Q is not a function of temperature with the NPO formulation. Plots of dissipation factor versus temperature, frequency, and
AC voltage for K1200 dielectric are presented in Figure 4.13.

Dielectric Constant as a Function of Applied DC Voltage.
Class 1 or NPO dielectrics are not polarized by applied DC voltages
and thus retain the same dielectric constant independent of applied
voltage. Class 2 dielectrics, however, show 10, 25, and even 75%
decreases in capacitance with high DC voltage levels (Figure 4.14).
Moreover, there is no predictable rate of return to the original
capacitance value as occurs with capacitance change due to increased
temperature. Polarized capacitors can be reformed by heating the
dielectric past the Curie point to remove polarization, after which
capacitance returns to its original value in a predictable manner.
NPO material is not affected by AC voltages, which do, however,
cause capacitance and dissipation increases in K1200 dielectrics.
Generally, the more available plate termination area, the higher
the AC current that may be carried. Thus high-capacitance units
can pass higher AC current levels than can smaller capacitors.

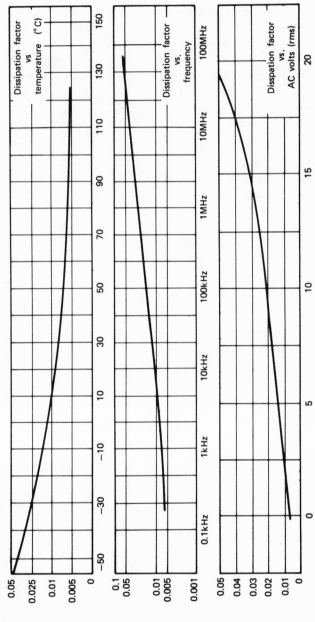

Figure 4.13. Dissipation factor versus temperature, frequency, and ac voltage, K1200 material. Courtesy of Monolithic Dielectrics, Inc.

71

In interpreting capacitance changes caused by the ac field, it is important to know the frequency and voltage used to measure the capacitance of high K ceramics.

Effects of Frequency on Capacitance Value. NPO materials, unaffected by frequencies to 10 MHz, show gradual increases in dissipation factor and gradual decreases in capacitance above this frequency. At 100 MHz dissipation is about 2% with capacitance at 93% of the 10-MHz value. A K1200 dielectric (Figure 4.14) begins to degrade at 1 MHz, where dissipation factor is about 3% and capacitance about 94% of its 100 kHz values. In both instances, electrical characteristics decrease as frequency is increased.

Self-inductance of a chip capacitor also limits useful frequency range and can be represented as a series LC circuit where L is largely composed of the inductance of the capacitor leads. This inductance, which is on the order of 10^{-8} or 10^{-9} H irrespective of capacitance value, may be reduced by shortening capacitor lead lengths, by using feed-through capacitors, or by shielding the capacitor.

Reliability. When properly attached to a substrate, a ceramic chip capacitor is reliable against changes due to moisture, shock, vibration, and normally encountered temperature extremes. In choosing a ceramic chip capacitor, one should consider that capacitor reliability is proportional to dielectric thickness, since capacitor failure is, in general, proportional to the cube of the electric field strength, which in turn is a function of the distance between capacitor plates. If enough volume is available, higher reliability can be insured by using a thicker ceramic. The single layer chip, then, is inherently more reliable than are multilayer types.

One must be careful to minimize the tarnishing of silver-containing end terminations, which also decreases reliability. Sulfur present in the air and within wrapping and labeling paper is the greatest contributor to this tarnishing. Labels should never be placed in capacitor containers. Care must be taken in handling chip capacitors, since finger prints oils will also decrease insulation resistance and dissipation factor. Chips should be handled with plastic tweezers and cleaned in alcohol or trichloroethylene. When soldering chips, minimizing conductor temperatures will reduce the tendency of silver end terminations to leach into the solder.

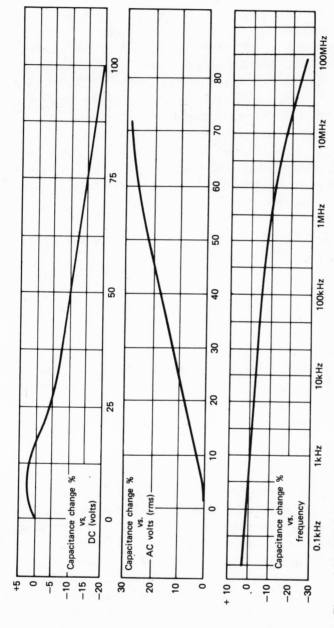

Figure 4.14. Capacitance change versus dc voltage, ac voltage, and frequency, K1200 material. Courtesy of Monolithic Dielectrics, Inc.

73

If a general-purpose 1200K ceramic has been heated above 115°C while soldering the chip or curing epoxy, capacitance can increase as much as 5%. Aging tends to bring this value back to its specified level. The life of a ceramic capacitor is affected by its applied voltage as well as the temperature at which it is used, since expected life varies inversely as the cube of the applied voltage. For example, a 0.002-mF ceramic capacitor with 10-V DC bias will have approximately eight times the expected life of another identical capacitor subjected to a 20-V DC bias. Aging effects due to heating during manufacture generally are not noticeable at the time the capacitor is delivered to the customer, because major aging effects occur within one day of the time of last heating.

Capacitors received from a vendor are usually subjected to incoming test. The correct sequence for this inspection is as follows: (1) test for capacitance and dissipation factor; (2) check insulation resistance; (3) test dielectric strength. If an attempt is made to check capacitance after insulation resistance and dielectric strength tests are made, the capacitance will measure low, an effect caused by applied voltage stress.

Summary NPO characteristic capacitors are much larger than K1200 or K6000 devices but smaller than equivalent mica or glass capacitors; they are available with capacities to 0.027 MF at tolerances of 1 to 20%; they are relatively expensive, costing about as much as K1200 capacitors with 40 times the capacitance; they have negligible dissipation factor; they show insulation resistance which increases with increasing temperature; they follow a linear ±10 ppm/°C capacitance variation with temperature; they possess a Curie point at a temperature high enough to be outside the range of interest; they show no noticeable effects due to aging or applied DC or AC voltage; and they have electrical parameters that do not begin to decrease at frequencies lower than 100 MHz.

W characteristic K1200 capacitors are small, equivalent to tantalum in lower capacitance ranges; they are available to 1.5 μF with tolerances of 5, 10, or 20%; they are inexpensive, but their cost increases with increasing capacitance; they have a dissipation factor that decreases with increasing temperature and insulation resistance that decreases with increasing temperature; they show a nonlinear ±10% capacitance change with temperature over the region of interest; with Curie point of 115°C, they have 1 or 2%/hour/decade aging affect; they produce noticeable capacitance changes due to either dc or ac applied voltages; and they show degradation of electrical parameters at 1 MHz or higher frequencies.

Tantalum capacitors, which have capacities between 0.33 μF at 35 V to 220 μF at 3 V, are available for bypass, coupling, or blocking applications. These are relatively small physically, yet they are expensive and do not exhibit the tolerance and stability characteristics of ceramic capacitors.

MICROCIRCUIT INDUCTORS

Although it is true that inductors and transformers cannot be used for monolithic microcircuits, this notion cannot be applied universally to hybrid thick or thin film devices. Certainly "chip" or screened and fired inductors in the nanohenry range can be included in UHF and microwave hybrids.

Minature toroidal or bar type transformers and inductors (Figure 4.15) also find application in hybrid microcircuit manu-

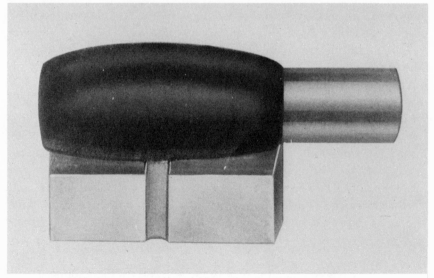

Figure 4.15.   Tunable microinductor. Courtesy of Piconics, Inc.

facture (7). Through the use of ferrite materials, it is possible to obtain inductance of 1000 μH within a volume of 250 $\times$ 250 $\times$ 70 mils. A coil with volume smaller than a transistor chip can deliver a useful Q of 15. Since most applications use loaded Q's of 2 to 10, this value is entirely satisfactory. Unfortunately, inductors are miniaturized at the cost of power handling ability. Although some microinductors will handle peak RF powers of 100 W, the important parameter is continuous power rather than peak power, since longterm heating effects are a major cause of component failure. Piconex, an important producer of small inductors, reports that continuous input powers of 2 W can be obtained with small inductors.

Fortunately, flux leakage can be controlled to insure a good tradeoff between Q degradation and weight reduction. Micromagnetic components can give weight and volume reductions on the order of 10 to 1000 times while reducing Q on the order of 2 to 1 or less! Since flux leakage is small, inductors of this size often can be used without heavy magnetic shielding.

Inductor miniaturization greatly improves reliability and vibrational resistance through reduced weight and lead length. Since the same ferrite materials are used, temperature coefficients for microinductors are identical to those found in larger units.

Small inductors for RF and IF applications must be tuneable (Figure 4.16) to compensate for the poor tolerances of associated fixed components. However, although toroids offer very high inductances for their volume, they do not allow much opportunity for tuning. Therefore solenoids and bobbins are usually employed. Miniature inductors find application in the UHF region (Figure 4.18), and low Q, high permeability, 80-mil diameter toroids, which may have inductances to 100 μH, can be used as pulse and broad band transformers. Solenoids and bobbins between 12 and 40 mils in diameter have inductances of 7 nH to 20 mH and possess the higher Q, lower permeability, restricted magnetic field, and high stability required for RF and IF amplifiers, discriminators, and modulators and similar circuitry. Various toroid, solenoids, and bobbin coils have characteristics that make them useful in the VHF and UHF regions to 3 GHz. Figure 4.17 is a comprehensive list of microinductor properties.

Alternatives to the use of inductors must be found for circuits for design inductances over 20 mH. One useful design alternative

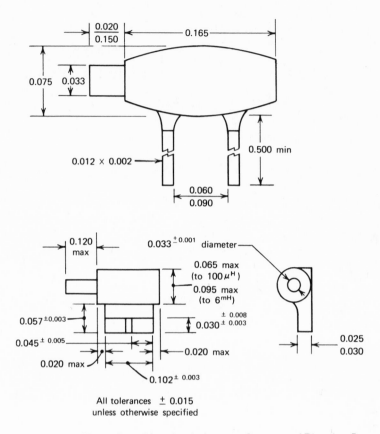

Figure 4.16. Two adjustable microinductors. Courtesy of Piconics, Inc.

is an *RC* active filter. *Q* multiplier circuitry, active filters, resonant semiconductors, and crystal filters can also simulate or replace inductors. Large passive *LC* filters are often replaceable by active *LC* filters.

Microinductors should be handled carefully with precision tweezers or a micromanipulator. Piconex has found that the largest single cause of breakage during assembly is soldering leads first and allowing the inductor to be entirely supported by the leads until bonded. This company suggests avoiding stress by fastening the body of the inductor with a drop of epoxy then soldering or bonding its leads, after which the inductor can be tuned.

PART NO PV SERIES	PART NO. G SERIES	RANGE in uh L max ±10%	L min ±10%	LEAD MATERIAL G SERIES ONLY	AT L MAX 10 MHZ Q Typ	Q Min	L uh	AT L MAX 30 MHZ Q Typ	Q Min	L uh	AT L 60 MHZ Q Typ	Q Min	L uh	AT L MAX 120 MHZ Q Typ	Q Min	L uh	At L Max 240 MHZ Q Typ	Q Min	L uh	At L Max Srf MHZ Typ	Srf MHZ Min	DCR Typ ohms	DCR Max ohms	Idc max ma +	Tc ppm/°C 50%
PV250K8I	G250K8I-40	.025	.018	Copper				60	51	.027	59	50	.034	75	64	.027	69	59	.025	1400	1050	.011	.014	250	100
PV250K8F	G250K8F-40	.025	.009	Copper				60	51	.022	68	58	.025	91	77	.021	11	9.4	.029	1400	1050	.007	.009	250	700
PV450K8I	G450K8I-40	.045	.025	Copper				56	48	.044	57	49	.046	73	62	.043	54	46	.050	1100	900	.100	.112	244	100
	G450K8I-30	.045	.025	Gold Ribbon				51	43	.030	64	54	.038	71	60	.036	50	43	.046	1100	900	.100	.112	244	100
PV450K8F	G450K8F-40	.045	.015	Copper				60	51	.044	60	51	.047	46	39	.042	19	16	.044	1070	803	.015	.020	250	700
PV560K8F	G560K8F-40	.056	.019	Copper				60	51	.058	62	53	.064	59	50	.058	21	18	.059	900	675	.012	.016	250	700
PV650K8F	G650K8F-40	.065	.022	Copper				44	37	.063	53	45	.058	76	65	.063	9.5	8.1	.065	880	660	.015	.020	250	700
PV650K3F	G650K3F-40	.065	.011	Copper	23	19	1.30	30	26	.070	20	17	.064	12	10	.067	7.5	6.4	.061	880	660	.036	.047	250	100
PV800K7I	G800K7I-40	.080	.032	Copper	32	27	.097	60	51	.078	57	49	.075	57	49	.080	27	23	.076	790	590	.023	.030	250	95
PV101K7I	G101K7I-20	.100	.040	Copper Ribbon	34	29	.116	60	51	.102	51	43	.112	63	54	.104	24	20	.098	735	550	.024	.031	250	95
PV101K6I	G101K6I-20	.100	.029	Copper Ribbon	35	30	.112	57	49	.100	42	36	.110	34	29	.102	18	15	.098	685	514	.026	.034	250	65
PV101K8F	G101K8F-20	.100	.025	Copper Ribbon	37	31	.117	60	51	.104	65	55	.099	42	36	.106	14	12	.101	690	520	.024	.031	250	700
PV101K8I	G101K8I-30	.100	.050	Gold Ribbon				54	48	.095	56	50	.111	60	54	.116	57	50	.106	780	650	.300	.390	140	100
PV121K8F	G121K8F-20	.120	.030	Copper Ribbon	34	29	.142	46	39	.119	34	29	.121	14	12	.127	6.4	5.4	.121	590	440	.059	.077	250	700
PV151K8I	G151K8I-30	.150	.075	Gold Ribbon				54	48	.155	59	53	.171	54	48	.211	40	35	.166	650	540	.300	.390	140	100
PV151K8I	G151K8I-20	.150	.075	Copper Ribbon	19	16	.172	41	35	.140	62	53	.121	48	41	.148	20	17	.148	650	540	.300	.390	140	100
PV151K6I	G151K6I-20	.150	.043	Copper Ribbon	23	20	.170	44	37	.140	47	40	.144	50	43	.139	22	19	.140	705	530	.095	.124	250	65
PV151K8F	G151K8F-20	.150	.038	Copper Ribbon	33	28	.158	53	45	.148	60	51	.142	50	43	.149	17	14	.152	575	430	.068	.070	250	700
PV181K6I	G181K6I-20	.180	.051	Copper Ribbon	25	21	.200	38	32	.169	37	31	.174	34	29	.166	20	17	.169	620	470	.120	.160	225	65
PV221K8I	G221K8I-20	.220	.110	Copper Ribbon	27	23	.236	47	40	.210	41	35	.209	39	33	.213	25	21	.240	570	450	.300	.390	140	100
PV221K8I	G221K8I-30	.220	.110	Gold Ribbon				53	47	.165	46	41	.232	44	39	.245	40	35	.211	570	450	.300	.390	140	100
PV221K7I	G221K7I-20	.220	.088	Copper Ribbon	25	21	.240	46	39	.230	40	34	.210	38	32	.220	19	16	.210	565	420	.150	.200	200	95

Figure 4.17. Electrical characteristics of small microinductors. Courtesy of Piconics, Inc.

PART NO. PV SERIES	PART NO. G SERIES	L max ±10%	L min ±10%	LEAD MATERIAL G SERIES ONLY	2.5 MHZ Q Typ	Q Min	L uh	10 MHZ Q Typ	Q Min	L uh	30 MHZ Q Typ	Q Min	L uh	60 MHZ Q Typ	Q Min	L uh	120 MHZ Q Typ	Q Min	L uh	240 MHZ Q Typ	Q Min	L uh	srf MHZ Typ	srf MHZ Min	DCR TYP ohms	DCR Max ohms	Idc max ma	Tc Dppm°C ±50%
PV221K6I	G221K6I-20	.220	.063	Copper Ribbon				25	21	.240	48	41	.260	53	45	.260	47	40	.260	18	15	.260	660	495	.14	.18	208	65
PV271K7I	G271K7I-20	.270	.110	Copper Ribbon				30	26	.270	43	37	.270	54	46	.260	52	44	.260	17	14	.280	535	400	.18	.23	183	95
PV331K8I	G331K8I-20	.330	.170	Copper Ribbon				32	27	.340	49	42	.320	61	52	.320	53	45	.310	14	12	.380	450	360	.30	.39	140	100
	G331K8I-30	.330	.170	Gold Ribbon							56	47	.340	64	54	.350	60	51	.330	32	27	.360	450	360	.30	.39	140	100
PV331K8F	G331K8F-20	.330	.083	Copper Ribbon				28	24	.350	38	32	.330	48	41	.310	40	34	.390	13	11	.360	380	290	.16	.21	194	700
PV391K8I	G391K8I-20	.390	.195	Copper Ribbon				28	24	.410	42	36	.380	58	49	.380	48	41	.390	40	34	.490	360	270	.20	.26	174	100
PV391K6I	G391K6I-20	.390	.111	Copper Ribbon				34	29	.390	43	37	.380	44	37	.370	16	14	.490	16	14	.490	382	290	.20	.26	174	65
PV471K7I	G471K7I-20	.470	.190	Copper Ribbon				33	28	.510	46	39	.510	65	55	.450	58	49	.450	12	10	.500	376	280	.20	.26	174	95
PV471K8F	G471K8F-20	.470	.120	Copper Ribbon				38	32	.470	48	41	.480	36	31	.490	19	16	.490	7.5	6.4	.570	370	300	.20	.26	174	700
	G471K8F-30	.470	.120	Gold Ribbon							46	41	.455	48	43	.531	47	42	.491	40	35	.451	370	300	.20	.26	174	700
PV561K8I	G561K8I-20	.560	.280	Copper Ribbon				27	23	.550	34	29	.550	53	45	.530	64	54	.540	44	37	.610	395	296	.50	.65	109	100
PV681K7I	G681K7I-20	.680	.270	Copper Ribbon				39	33	.660	43	37	.680	60	51	.660	47	40	.670	31	26	.920	310	230	.60	.78	100	95
PV681K6I	G681K6I-20	.680	.195	Copper Ribbon				40	34	.660	38	32	.670	39	33	.640	21	18	.660	10	8.5	.770	324	240	.50	.65	109	65
PV681K8F	G681K8F-20	.680	.170	Copper Ribbon							37	33	.791	35	31	.931	30	27	.866				227	180	.30	.39	140	700
	G681K8F-30	.680	.170	Gold Ribbon				35	30	.660	38	32	.680	44	37	.520	19	16	.670				227	180	.50	.65	109	700
PV821K7I	G821K7I-20	.820	.330	Copper Ribbon				36	31	.770	36	31	.820	58	49	.770	32	27	.810				324	240	.50	.65	109	95
PV102K3F	G102K3F-20	1.00	0.11	Copper Ribbon	25	21	1.35	23	20	.970	12	10	.950	7.1	6.0	.950	4.1	3.5	.910				270	200	.20	.26	174	100
PV102K7I	G102K7I-20	1.00	0.40	Copper Ribbon	23	20	1.06	30	26	.980	43	37	.970	48	41	1.00	34	29	1.15				196	147	.40	.52	123	95
PV102K6I	G102K6I-20	1.00	0.29	Copper Ribbon	32	27	1.05	49	42	.980	58	49	.960	47	40	1.01	22	19	1.06				270	200	.30	.39	142	65
PV102K8F	G102K8F-30	1.00	0.25	Gold Ribbon							50	45	.950	59	50	1.00	46	41	1.20				220	175	.40	.52	123	700
PV122K6I	G122K6I-20	1.20	0.34	Copper Ribbon	25	21	1.36	43	37	1.19	46	39	1.13	37	33	1.18	22	19	1.24				285	214	.50	.65	109	65
PV152K6I	G152K6I-20	1.50	0.43	Copper Ribbon	29	25	1.53	50	43	1.45	52	44	1.63	36	31	1.47							220	165	.60	.78	100	65
PV152K3I	G152K3I-20	1.50	0.38	Copper Ribbon	31	26	1.53	29	25	1.48	21	18	1.43	12	10	1.50							185	139	.30	.39	142	150

Figure 4.17. Continued

PART NO. PV SERIES	PART NO. G SERIES	RANGE in uh L max +10%	L min +10%	LEAD MATERIAL G SERIES ONLY	AT L MAX 1 MHZ Q TYP	Q Min	L uh	AT L MAX 2.5 MHZ Q TYP	Q Min	L uh	AT L MAX 10 MHZ Q TYP	Q Min	L uh	AT L MAX 30 MHZ Q TYP	Q Min	L uh	AT L MAX 60 MHZ Q TYP	Q Min	L uh	AT L MAX 120 MHZ Q TYP	Q Min	L uh	AT L MAX srf MHZ TYP	srf MHZ Min	DCR TYP ohms	DCR Max ohms	Idc max ma *	Tc ppm/°C ± 50%
PV152K8F	G152K8F-30	1.50	0.38	Gold Ribbon										53	47	1.63	49	44	1.60	24	21	1.66	210	165	.40	.52	123	700
PV182K6I	G182K6I-20	1.80	0.52	Copper Ribbon				29	25	1.84	50	43	1.74	52	44	1.96	36	31	1.77				200	150	.72	.94	91	65
PV222K7I	G222K7I-20	2.20	0.88	Copper Ribbon				31	26	2.20	42	36	1.77	36	31	2.60	12	10	5.60				70	52	.90	1.2	82	95
PV222K8F	G222K8F-30	2.20	0.55	Gold Ribbon							35	30	2.20	36	32	2.50	47	42	2.50				110	85	.40	.52	123	700
PV272K7I	G272K7I-20	2.70	1.10	Copper Ribbon				31	26	2.60	40	34	2.70	36	32	3.00							66	49	.90	1.2	82	95
	G332K8F-20	3.30	0.83	Copper Ribbon				34	29	3.00	30	26	3.10	29	25	2.90							57	45	1.5	.65	109	700
PV332K8F	G332K8F-30	3.30	0.83	Gold Ribbon				35	30	3.20	31	25	3.40	30	27	3.30							57	45	.50	.65	109	700
PV332K6I	G332K6I-20	3.30	0.94	Copper Ribbon				33	28	3.20	29	25	3.40	28	24	3.30							105	79	1.5	2.0	64	65

Figure 4.17. Continued

80

Figure 4.17. Continued

PART NO. FV SERIES	PART NO. G SERIES	RANGE L max +10%	L min +10%	LEAD MATERIAL G SERIES ONLY	AT L MAX 1 MHZ Q Typ	Q Min	L uh	AT L MAX 2.5 MHZ Q Typ	Q Min	L uh	AT L MAX 10 MHZ Q Typ	Q Min	L uh	AT L MAX 30 MHZ Q Typ	Q Min	L uh	AT L MAX 60 MHZ Q Typ	Q Min	L uh	AT L MAX 120 MHZ Q Typ	Q Min	L uh	AT L MAX srf MHZ Typ	srf MHZ Min	DCR Typ ohms	DCR Max ohms	Idc max mA *	Tc PPM/°C ± 50%
PV332K3F	G332K3F-20	3.30	0.37	Copper Ribbon				22	19	3.30	23	20	3.30	11	9.3	3.40							78	59	.30	.39	142	100
PV392K6I	G392K6I-20	3.90	1.11	Copper Ribbon				24	20	3.80	48	41	4.00	37	32	3.90							90	68	1.6	2.1	62	65
PV472K6I	G472K6I-20	4.70	1.34	Copper Ribbon				31	26	4.20	37	31	4.40	25	21	4.70							65	49	1.7	2.2	60	65
PV472K8F	G472K8F-30	4.70	1.18	Gold Ribbon				45	38	4.20	46	39	5.00	23	20	6.00							54	43	.70	.91	9?	700
PV582K3F	G582K3F-20	5.80	0.65	Copper Ribbon				40	34	5.40	37	32	6.10	14	12	5.40							120	90	1.2	1.6	71	100
PV682K3F	G682K3F-20	6.80	0.76	Copper Ribbon				37	31	6.30	25	21	6.30	11	9.3	6.80							70	53	.90	1.2	82	100
	G682K8F-20	6.80	1.70	Copper Ribbon				50	43	6.80	59	50	7.19	30	27	7.20							51	40	1.0	1.3	78	700
PV682K8F	G682K8F-30	6.80	1.70	Gold Ribbon				50	43	6.90	60	51	7.20	30	27	7.20							51	40	1.0	1.3	78	700
PV822K6I	G822K6I-20	8.20	2.34	Copper Ribbon	20	17	8.00	42	36	8.00	53	45	8.80										36	27	2.1	2.7	54	65
PV103K6I	G103K6I-20	10.0	2.85	Copper Ribbon	18	15	9.40	42	36	9.40	49	42	10.9										25	19	3.0	3.9	45	65
PV103K3I	G103K3I-20	10.0	2.50	Copper Ribbon	22	19	9.80	41	35	10.0	34	29	11.1										25	19	2.4	3.1	50	150
PV103K3F	G103K3F-20	10.0	1.11	Copper Ribbon	31	27	10.0	49	42	10.0	39	33	10.0										51	38	1.8	2.3	58	100
PV103K8F	G103K8F-30	10.0	2.50	Gold Ribbon				51	43	10.0	31	26	12.0										28	22	1.3	1.7	68	700
PV123K3F	G123K3F-20	12.0	1.33	Copper Ribbon	31	27	11.9	46	39	11.9	36	32	12.2										35	26	2.2	2.9	53	100
PV153K3I	G153K3I-20	15.0	3.80	Copper Ribbon	28	24	14.9	43	37	15.0	29	25	17.0										23	17	2.8	3.7	47	150

Figure 4.17. Continued

PART NO. PV SERIES	PART NO. G SERIES	L max −10%	L min ±10%	LEAD MATERIAL G SERIES ONLY	250 KHZ Q Typ	250 KHZ Min	250 KHZ L uh	.5 MHZ Q Typ	.5 MHZ Min	.5 MHZ L uh	1 MHZ Q Typ	1 MHZ Min	1 MHZ L uh	2.5 MHZ Q Typ	2.5 MHZ Min	2.5 MHZ L uh	10 MHZ Q Typ	10 MHZ L uh	srf MHZ TYP	srf MHZ Min	DCR TYP ohms	DCR Max ohms	Idc max ma	Tc ppm/°C ±50%
PV153K3F	G153K3F-20	15.0	1.67	Copper Ribbon							32	27	14.8	41	35	15.0	29	15.0	35	26	1.8	2.3	58	100
PV153K8F	G153K8F-30	15.0	3.75	Gold Ribbon										47	40	15.0	30	20.0	27	21	1.8	2.3	58	700
PV183K3F	G183K3F-20	18.0	2.00	Copper Ribbon							29	25	17.7	40	34	17.7	26	19.0	26	20	2.9	3.8	46	100
PV243K3I	G243K3I-20	24.0	6.00	Copper Ribbon							28	24	23.0	44	37	24.0			12.5	9.4	4.5	5.9	37	150
PV333K3I	G333K3I-20	33.0	8.25	Copper Ribbon							32	27	32.0	43	36	33.0			14.7	11.0	9.3	12.0	27	150
PV433K3F	G433K3F-20	43.0	4.80	Copper Ribbon				25	21	38.0	40	34	39.0	49	42	43.0			8.5	6.4	4.0	5.2	41	100
PV563K3F	G563K3F-20	56.0	6.20	Copper Ribbon				26	22	48.0	43	37	49.0	47	40	56.0			8.4	6.3	5.2	6.8	36	100
PV753K3I	G753K3I-20	75.0	18.8	Copper Ribbon				25	21	63.0	42	36	65.0	49	42	75.0			7.8	5.9	21	27	18	150
PV104K3F	G104K3F-20	100.0	11.1	Copper Ribbon				37	32	85.0	49	42	84.0	39	33	99.0			8.4	6.3	9.3	12	27	100
PV134K3F	G134K3F-20	130.0	14.5	Copper Ribbon				33	28	127	44	37	133	35	30	153			5.5	4.1	12	16	24	100
PV154K3F	G154K3F-20	150.0	16.7	Copper Ribbon				33	28	146	44	37	153	35	30	176			5.2	3.9	14	18	22	100
PV474K1F	G474K1F-20	470.0	49.5	Copper Ribbon				34	29	430	34	29	500						2.8	2.1	23	30	16	500
PV684K1F	G684K1F-20	680.0	72.0	Copper Ribbon				39	33	620	35	30	745						2.8	2.1	26	34	15	500
PV105K1F	G105K1F-20	1000	105	Copper Ribbon	26	22	990	35	30	1080	29	25	1370						1.5	1.1	56	73	11	500
PV285K1F	G285K1F-20	2800	295	Copper Ribbon	35	30	2770	35	30	3540									1.1	.83	86	112	8.8	500
PV335K1F	G335K1F-20	3300	350	Copper Ribbon	32	27	3270	32	27	4200									1.04	.78	98	128	8.3	500
PV585K1F	G585K1F-20	5800	610	Copper Ribbon	27	23	4800	27	23	5800									0.75	.56	160	208	6.5	500

NOTE: Where columns are blank, the reading could not be taken. At high frequency, no reading indicates the measurement is above self resonance. At low frequency no reading indicates that the measurement is beyond the limits of the test equipment.

*These rated maximum currents are one-half the amount that will cause a 30°C temperature rise unless the rated current exceeds 250ma. Then a rating of 250ma is used.

MATERIAL LEAD CODE

—02   .002 x .012 KOVAR RIBBON

—20   .002 x .012 COPPER RIBBON

—30   .002 x .010 GOLD RIBBON

—40   .007 TO .018 DIA. COPPER WIRE

Figure 4.17.   Continued

MCH5800
thru
MCH5805

MOTOROLA
Semiconductors
BOX 20912 • PHOENIX, ARIZONA 85036

Microcircuit Components

UNENCAPSULATED THIN-FILM INDUCTORS

28 to 230 nH

AUGUST 1969 – DS 9140

THIN-FILM CHIP INDUCTORS

. . . designed for use in UHF and microwave hybrid circuits for tuning and biasing applications.

- High Q
- Uniform Inductance
- High Self-Resonant Frequency
- Small Physical Size

DEVICE OUTLINE AND DIMENSIONS

Figure 4.18. UHF unencapsulated thin film inductor. Courtesy of Motorola Semiconductor Products, Inc.

84

MAXIMUM RATINGS (T_A = +25°C unless otherwise noted)

Rating	Symbol	Value	Unit
Operating Current	$I_{C(max)}$	250	mA
Peak Device Temperature (relative to mounting)	T_{pk}	300	°C

Type No.	(Inches ± 10 m)		
	L	W	T
MCH5800	0.190	0.190	0.010
MCH5801	0.192	0.192	0.010
MCH5802	0.220	0.220	0.010
MCH5803	0.225	0.225	0.010
MCH5804	0.233	0.233	0.010
MCH5805	0.240	0.240	0.010

TYPICAL ELECTRICAL CHARACTERISTICS (T_A = +25°C unless otherwise noted)

Characteristic	Symbol	MCH5800	MCH5801	MCH5802	MCH5803	MCH5804	MCH5805	Unit
Inductance	L_S	28	35	115	150	195	230	nH
Figure of Merit	Q	19	21	27	25	29	30	—
Test Frequency	f	200	200	200	100	100	100	MHz
Self-Resonant Frequency	f_R	1800	1700	1030	920	810	740	MHz
Series Resistance	R_S	1.8	2.1	5.5	4.0	4.5	5.0	ohms
Stray Capacitance	C_S	0.27	0.25	0.21	0.20	0.20	0.20	pF

Figure 4.18. Continued

APPLICATIONS INFORMATION

1. Equivalent Circuit of Thin-Film Inductor:

Figure 1 is an equivalent circuit of the thin-film inductor.

C_S — Distributed capacitance
R_S — Resistance of inductor (function of frequency)
L_S — Actual inductance (low-frequency)

2. Equivalent Series Circuit of Thin-Film Inductor:

The presence of distributed capacitance affects the actual series resistance and inductance of the coil resulting in the equivalent series circuit shown in Figure 2.

The equivalent inductance and resistance is related to the actual inductance and resistance according to the equations below:

$$L_{equiv} = \frac{L_S}{1-K^2}; \text{ where } K = \frac{f}{f_R},$$

f is the frequency of interest, and f_R is the frequency at which the coil inductance is resonant with distributed capacitance (C_S).

$$R_{equiv} = \frac{R_S}{(1-K^2)^2}$$

3. Self-Resonant Frequency:

The self-resonant frequency of the thin-film inductor is:

$$f_R = \frac{1}{2\pi\sqrt{L_S C_S}}$$

C_S (pF) varies with the number of turns and is found to be $C_S = 0.15 + \frac{0.42}{N}$; where N is the number of turns.

4. Inductance as Function of Turns:

A plot of inductance (actual) versus the number of turns is shown in Figure 3.

5. Spiral Inductor Figure of Merit, Q:

The Q of a spiral inductor can be shown as:

$$Q = \frac{\omega L_{equiv}}{R_{equiv}}$$

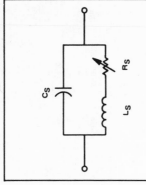

FIGURE 1 – EQUIVALENT CIRCUIT

C_S

L_S R_S

Measurements may be made using a Boonton Radio Model 250-A RX Meter or equivalent.

FIGURE 2 – EQUIVALENT SERIES CIRCUIT

L_{equiv} R_{equiv}

For design purposes at frequencies approximately $\frac{f_R}{8}$ or lower, C_S can be neglected, therefore Q can be written as $Q = \frac{\omega L_S}{R_S}$.

6. General Formulas Involving Q:

(A) $R_P = R_{equiv}(1 + Q^2)$; where
R_P = equivalent parallel resistance

(B) $L_P = L_{equiv} \cdot \frac{1 + Q^2}{Q^2}$; where L_P = equivalent parallel inductance

(C) $Q = \frac{f_c}{f1 - f2}$; where f1 and f2 are half power points and f_c is center frequency

(D) $Q = \frac{R_P}{\omega L_P}$

NOTE: These miniature thin-film inductors are fabricated on a 0.010" thick alumina substrate using copper as the primary material. Motorola uses a unique photo-resist processing technique which makes possible conductor thicknesses not previously achievable in this small geometry.

We recommend that the devices be mounted using a nonconductive organic adhesive such as epoxy. Wire bonds can be made using parallel gap welding, soldering, ultrasonic, or thermo-compression bonding techniques.

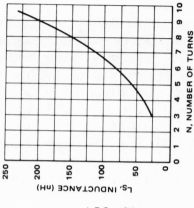

FIGURE 3 — TYPICAL ACTUAL
INDUCTANCE per NUMBER of TURNS

⊛ MOTOROLA Semiconductor Products Inc.

BOX 20912 • PHOENIX, ARIZONA 85036 • A SUBSIDIARY OF MOTOROLA INC.

508A PRINTED IN USA 8-69 IMPERIAL LITHO B12463   15M

DS 9140

Figure 4.18. Continued

SEMICONDUCTOR CHIPS

A major advantage of hybrid thick and thin film microcircuits over monolithics is that they can employ individually selected active devices (8–13), such as those shown in Figures 4.19 to 4.22. It is interesting to note that even monolithic integrated circuits such as the operational amplifier of Figure 4.22 can be included within the hybrid. Solid-state elements for hybrid applications are commonly mounted in a variety of micropackages, including the three-ribbon-leaded T package, LIDs or leadless inverted devices, and flip channel packages.

Active device selection is based on physical size limitations, electronic design parameters, degree of reliability required, electrical tolerances allowable, and finally the overall package requirements of the device. Form-sliced, chipped or mounted-electrical character-

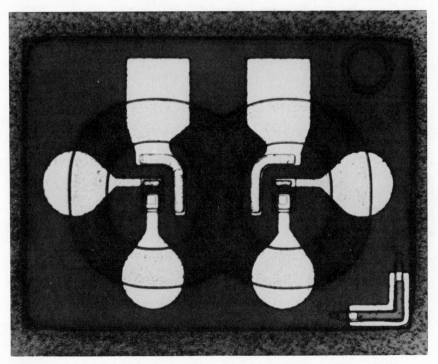

Figure 4.19. NPN dual transistor chip—monolithic. Courtesy of Intersil, Inc.

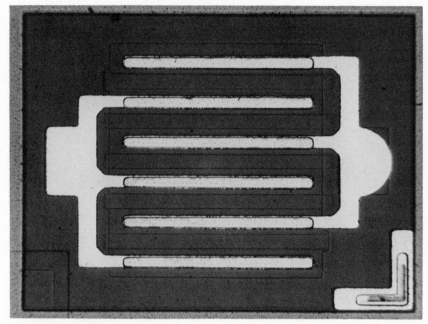

Figure 4.20. 2N4867 N channel J FET chip for low-noise audio applications. Courtesy of Intersil, Inc.

istics, handling procedures required for chip slices, and physical size all are important in procuring semiconductors.

Miniature chip semiconductors are mounted to the substrate by chip and wire bonding, nailhead and wedge bonding, thermocompression bonding, ultrasonic bonding, parallel gap soldering or welding, reflow soldering, a variety of backbonding techniques, including solder eutectic and adhesive methods, laser and electron beam welding, and flip chip face bonding. In addition, there are methods especially devised for specific package types, including beam leads, ceramic flip chips, LIDs and channels, recessed chip and evaporated connections, STD, spider, and SLT.

Chip Data Sheets. Figures 4.23 and 4.24 represent typical data sheets for solid-state chip components. Consider, for example, the data sheets for Motorola unencapsulated silicon bipolar transistors. These show pertinent electrical characteristics normally associated with transistor data sheets but omit those concerned with power

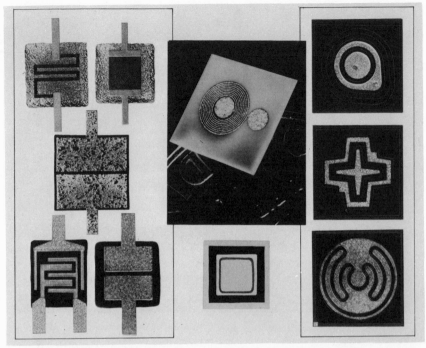

Figure 4.21. Chip semiconductors and passives. Courtesy of Motorola Semiconductor Products, Inc.

dissipation or thermal resistance. Since these depend largely on the mounting technique employed by the user of the device, the manufacturer cannot specify them accurately.

Semiconductor Procurement and Testing. Usually a large number of active device or diode chips are simultaneously manufactured on one semiconductor slice. One chip on each slice is designated the test chip and is given a full battery of electrical tests. If it does not pass, the entire slice can be rejected. Slices that pass an early examination are given a further sequence of tests including visual inspection and sample probing of DC parameters. Again, if a large percentage of the sample proves faulty the entire slice is rejected. Otherwise, all of the units on the slice are then probed and bad chips are marked with a spot of red dye. Some manufacturers use several dyes to specify different tolerances. After a complete probing the

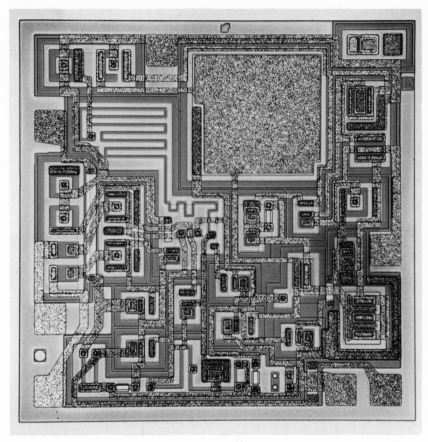

Figure 4.22. Operational amplifier μA741 due for thick film application. Courtesy of Fairchild Semiconductor.

slice is cut into chips and a sample group of these is mounted on standard transistor headers. A complete set of low- and high-frequency tests then can be performed to check agreement with published specifications.

When choosing semiconductor chips, it is advantageous to select standard off-the-shelf semiconductors, which are usually available in ample supply at reasonable prices and are of established quality. It is important to specify only the electrical characteristics and tolerances needed, since overspecification leads to increased costs and possible delivery delays while adding nothing of significance to the

STANDARD CHIP PROCESSING

The transistor and small-signal diode "chips" in Motorola's Standard Microcircuit Components line are produced on the same well-proven production lines that provide Motorola's standard encapsulated devices. They are subjected to the same rigid in-process controls used to insure the reliability and performance of the eventual packaged components. In fact, as shown in the flow chart below, all wafer processing is completed before the wafers are assigned either for subsequent encapsulation or for additional special testing and handling involved in selling unencapsulated components.

As with standard encapsulated products, the entire test and inspection sequence for chips is under the auspices of the Quality Control Department, providing independent quality assurance completely disassociated from production control.

CHIP PROCESSING AND QUALITY CONTROL SEQUENCE

WAFERS ASSIGNED TO SPECIAL ORDERS AND
PRODUCTION OF ENCAPSULATED DEVICES

Figure 4.23.   Transistor chip data and general chip information.   Courtesy of Motorola Semiconductor Products, Inc.

92

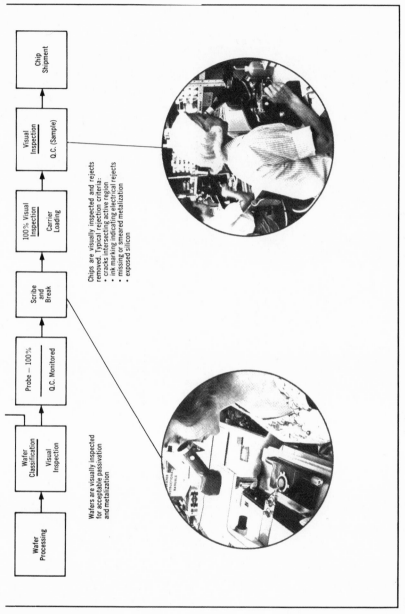

Wafer
Processing

Wafer
Classification
Visual
Inspection

Probe — 100%
Q.C. Monitored

Scribe
and
Break

100% Visual
Inspection
Carrier
Loading

Visual
Inspection
Q.C. (Sample)

Chip
Shipment

Wafers are visually inspected
for acceptable passivation
and metalization

Chips are visually inspected and rejects
removed. Typical rejection criteria:
• cracks intersecting active region
• ink marking indicating electrical rejects
• missing or smeared metalization
• exposed silicon

Figure 4.23. Continued

GENERAL INFORMATION

NON-STANDARD CHIP PROCESSING

The standard unencapsulated semiconductors described in the following sections meet a wide variety of application requirements. Nevertheless, there may be occasions when a designer can benefit from a non-standard device for a specific circuit. To satisfy these requirements, almost any device from Motorola's broad line of conventional packaged semiconductors may be obtained on a specially negotiated basis. Moreover, though the electrical specifications of these special chips are limited by certain test limitations, the customer can negotiate additional tests. Please contact your Motorola sales representative for more information.

On special order, Motorola transistors other than those listed in data sheets may be obtained in both wafer and chip form. The following tables list test limitations for these devices. The tests indicated can be made on a 100% basis. The tests can also be negotiated on a sampling basis.

Frequency and switching performance correspond to the inherent capability of a particular product line; dynamic specifications cannot be obtained by probing a chip. Such parameters are measured with the chip sealed in a standard encapsulated package and the resulting measurement includes the package parasitics.

TABLE I – Electrical Test Capability for 100% Wafer Probing of "Special" Unencapsulated Small-Signal and RF Transistors

(Tests are limited to 3 of the 4 indicated breakdown voltages)

Parameter*	Test Condition	Limits
BVEBO	$I_C = 10 \, \mu Adc$	10 Vdc
BVCBO	$I_C = 100 \, \mu Adc$	200 Vdc
BVCES	$I_C = 100 \, \mu Adc$	200 Vdc
BVCEO	$I_C = 10 \, mAdc$	200 Vdc
hFE**	$I_C = 1 \, mAdc$ to 150 mAdc $V_{CE} = 0$ to 20 Vdc	**

*In addition to the listed parameters, leakage current may be probed to limits no tighter than 20 nAdc.

**Choice of one hFE test to a maximum and a minimum limit or two hFE tests, each to a maximum or minimum limit.

TABLE II – Electrical Test Capability for 100% Wafer Probing of "Special" Unencapsulated Power Transistors

Parameter	Test Condition	Limits
BVEBO	$10 \, \mu Adc$-10 mA	30 Vdc
BVCBO	$50 \, \mu Adc$-5.0 mAdc	500 Vdc @ 1.0 mAdc
BVCES	$50 \, \mu Adc$-1.0 mAdc	500 Vdc @ 1.0 mAdc
BVCEO	1.0 mAdc-100 mAdc	500 Vdc @ 1.0 mAdc
hFE	$I_C = 50 \, mAdc$-1.0 Adc $V_{CE} = 1.0$-20 Vdc	—

Minimum leakage currents will be the same as the minimum currents listed for the breakdown voltages above. On high voltage material (100 Vdc) we convert the breakdown voltage to leakage currents for test purposes. hFE is test equipment limited, higher currents are correlated.

Figure 4.23. Continued

95

VISUAL INSPECTION

DEFINITION OF TERMS

Emitter-Base and Collector-Base Junctions. *The region where the base and collector, and the emitter and base meet. These junctions will be defined on the surface of the chip as an oxide step.*

Diffusion Window. *The opening etched through the oxide to permit the diffusion of the emitter and base.*

Active Junction. *A change in 'N' type to 'P' type doping or conversely, by a diffusion step. On discrete transistors there are 2 active junctions, the collector-base junction and the emitter-base junction.*

The Pre-Ohmic Window. *The opening etched through the oxide for metalization contact to the emitter and base regions.*

Pre-Ohmic Alignment. *The positioning of the oxide opening into which the metalization is placed.*

Passivated Region. *Any region covered by glass (Si O2), nitride, or other protective dielectric.*

Expanded Contact. *Any pattern that has metalization crossing a diffused junction.*

Attached Foreign Material. *A foreign substance that cannot be removed when subjected to a nominal gas flow. Lint, silicon dust, etc. are not considered attached since they can be removed after die mount.*

INSPECTION CRITERIA

Visual inspection is performed with a microscope using 40X-80X magnification for Silicon-Power Chips and 100X-125X for other devices.

SCRIBING DEFECTS

Excess Chip. A chip shall be rejected if a portion of an adjacent chip with metalization is still attached to subject chip.

Scribe Line Limits. A chip shall be rejected if a scribe line touches or crosses an active junction area or a metalized region.

FOREIGN MATERIAL DEFECTS

Bridged-Across Metal. A chip shall be rejected when attached foreign material bridges across normally separated metalized areas.

Particle Size Inside Active Area. A chip shall be rejected when attached foreign material greater than 2 mils. is found inside collector-base junction or on the emitter-base bonding pads.

MECHANICAL DEFECTS

Inspect each chip to insure there are no cracks or breaks that:

Non-Expanded Contacts

(a) Touch the collector-base junction (NPN).

(b) Extend through the annular ring (PNP).

Expanded Contacts

(a) Touch the collector-base junction (NPN).

(b) Extend through the annular ring (PNP).

(c) Extend under any metalized bonding pad.

Inspect each chip to insure there are no cracks greater than one mil. in length in a passivated region and extending toward an active area. (Does not apply to Silicon Power devices.)

ALIGNMENT DEFECTS

Pre-Ohmic Alignment. The chip shall not contain pre-ohmic windows that cross the edge of a diffusion window.

Diffusion Window Alignment. No diffusion window shall touch another diffusion window.

Metalization Alignment. The metalization must be aligned so that at least 50% of the pre-ohmic window is covered with metalization.

OXIDE DEFECTS

Exposed Silicon on Junction. A chip shall be rejected if exposed silicon touches or crosses the collector-base junction or the emitter-base junction.

Exposed Silicon Touching Metal. A chip shall be rejected if exposed silicon touches or extends under the bonding pad metalization. (Expanded contacts only.)

Oxide Defect in Active Area. A chip shall be rejected if an oxide defect occurs inside or on the collector-base junction with a major dimension greater than 1 mil. (Does not apply to Silicon Power Devices.)

Oxide Defect Crossing or Touching. A chip shall be rejected if gross oxide defects, evidenced by alternately colored bands (rainbow effect), emit from two separate ohmic contacts and either touch or cross each other, or cross the collector-base junction.

Figure 4.23. Continued

97

INSPECTION CRITERIA (continued)

Oxide Defect Under Bonding Pads. A chip shall be rejected if an oxide defect extends under 25% of the bonding pad.

METALIZATION DEFECTS

Expanded Contacts (finger geometries).

Missing Metalization on Bonding Pads. A chip shall be rejected when 25% of the metalization is missing from a bonding pad.

Metalization Width at Oxide Step. Any chip shall be rejected if the metalization width of any finger is reduced greater than 25% at any oxide step. 75% of the metal width must remain.

Metalization Width In First 50 Percent of Finger. A chip shall be rejected if the finger metalization is narrower than 50% of its original design width or if the finger width is reduced greater than 50% due to a severe scratch or void in the first 50% of the finger. A severe scratch is one which exposes the underlying surface.

Fingers Isolated or Missing. A chip shall be rejected if any finger is not 100% continuous over the first 50% of the finger (from the bonding pad).

Bubbled Metalization. A chip shall be rejected if it exhibits any bubbled metalization on a bonding pad.

any corroded metal. Corrosion is a chemical reaction or process causing abnormalities in the metalization. A rough metalization surface is not to be considered corrosion.

Non-Expanded Contacts

Missing Metalization. A chip shall be rejected when more than 25% of the metalization is missing from a bonding pad.

Lifted Metalization. A chip shall be rejected if it exhibits any lifted metalization. Slight undercutting causing a lifted appearance is not cause for rejection.

Bubbled Metalization. A chip shall be rejected if it exhibits any bubbled metalization on a bonding pad.

Bridged Metalization. A chip shall be rejected for bridged metal shorting any two normally separated metalized areas.

Narrow Metal Widths In Relation To Design Width. A chip shall be rejected if the metalization is narrower than 50% of its original design width.

Metal Corrosion. A chip shall be rejected if it exhibits any corroded metal. Corrosion is a chemical reaction or process causing abnormalities in the metalization. A rough metalization surface is not to be considered corrosion.

Lifted Metalization. A chip shall be rejected if it exhibits any lifted metalization. Slight undercutting causing a lifted appearance is not cause for rejection.

Bridged Metalization. A chip shall be rejected for bridged metal shorting any two normally separated metalized areas.

Metal Corrosion. A chip shall be rejected if it exhibits

METALIZED ANNULAR RING

Missing Metalization. A chip shall be rejected when a metalized annular ring is not 100% continuous.

Bridged Metalization. A chip shall be rejected for bridged metal shorting the metalized annular ring with any other metalized area.

RECOMMENDED INCOMING INSPECTION PROCEDURES

Motorola assures that the devices will meet the customers' incoming visual inspection when inspected to the visual criteria and LTPD limits specified in the data sheet. Inspection must be performed at the power and magnification indicated. Motorola guarantees dc parameters to LTPD limits specified in the data sheet.

Returned Components

It is suggested that the customer perform incoming inspection in the following sequence:

1. Visual

2. Test dc electrical parameters

A. If the lot fails visual inspection, containers must be closed and secured and the entire lot returned to Motorola with a detailed inspection report. In no case will Motorola accept rejected material that the customer has inspected 100%.

B. After the lot has passed incoming visual inspection, samples are selected and subjected to electrical tests of the dc parameters. If samples do not pass the electrical tests, they shall be packaged separately and identified with all the information from the original package of chips. The shipping container must be closed and secured. The entire lot together with the test samples and a detailed inspection report shall be returned to Motorola. In no case will Motorola accept rejected material that the customer has inspected 100%.

Figure 4.23. Continued

SHIPPING METHODS
FOR MICROCIRCUIT COMPONENTS

HANDLING PRECAUTIONS

Standard microcircuit components listed in the data sheets in this catalog are passivated devices, as are most special selections. However, many other unpackaged components, such as high power thyristors, silicon mesa power transistors and germanium power transistors, require special handling. Consequently their parameters cannot be guaranteed.

For passivated devices, although the care and handling of unencapsulated semiconductors often require precautions outside the experience of many equipment manufacturers, Motorola warrants that such devices meet or exceed the published (or negotiated) specifications, provided three basic requirements are met in the customer's establishment.

1. Such devices are stored in an environment of no more than 30% relative humidity.

2. Devices are processed in a non-inert atmosphere not exceeding 100°C, or in an inert atmosphere not exceeding 400°C.

3. Processing equipment conforms to the minimum standards of equipment normally employed by semiconductor manufacturers.

Moreover, Motorola's engineering staff is available for consultation in the event of correlation or processing problems encountered in the use of Motorola semiconductor chips. For assistance of this nature, please contact your nearest Motorola sales representative.

STANDARD CARRIER PACKAGES

To accommodate customers with both small and large quantity requirements, Motorola supplies microcircuit components in two standard carriers, the Deka-Pak and the Multi-Pak. These carriers are shown in Figures 1 and 2. Both contain individual compartments to simplify user inventory recordkeeping and to protect the chips during storage.

The Deka-Pak holds 10 small-signal chips, and is ideal for prototype development.

The Multi-Pak is excellent for production use. Two versions, both 2 inches square, are available. One holds 400 small-signal chips, and the other is designed for 100 large chips such as power transistors.

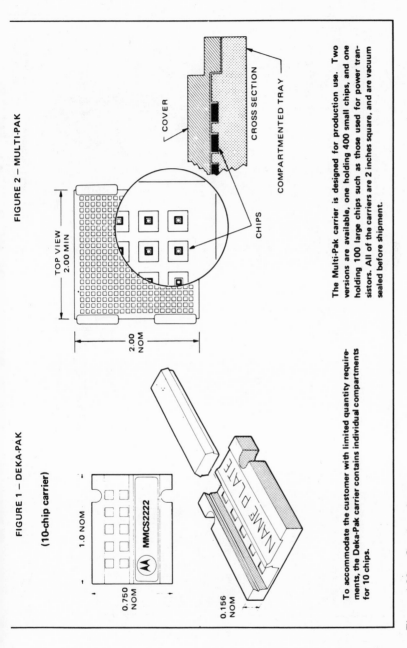

FIGURE 1 – DEKA-PAK

(10-chip carrier)

1.0 NOM

0.750 NOM

MMCS2222

0.156 NOM

NAME PLATE

FIGURE 2 – MULTI-PAK

TOP VIEW 2.00 MIN

2.00 NOM

COVER

CROSS SECTION

COMPARTMENTED TRAY

CHIPS

To accommodate the customer with limited quantity requirements, the Deka-Pak carrier contains individual compartments for 10 chips.

The Multi-Pak carrier is designed for production use. Two versions are available, one holding 400 small chips, and one holding 100 large chips such as those used for power transistors. All of the carriers are 2 inches square, and are vacuum sealed before shipment.

Figure 4.23.   Continued

101

SHIPPING METHODS
OPTIONAL SHIPPING METHODS

CHIP OPTIONS

For large quantity use, or special applications, shipping methods other than the standard Deka-Pak or Multi-Pak may be desired. Various packaging and shipping options are available on a negotiated basis. For more information on these options, please contact your Motorola sales representative.

FIGURE 3 – K-PAK (1000-CHIP CARRIER)

TOP VIEW

COMPARTMENTED TRAY

GLASS COVER

CHIPS

CROSS SECTION

This carrier holds 1000 chips. It is designed with individual compartments for each chip. The chips are placed in the

TABLE I – Specification Options

CHIPS	Shipping Options
1. 100% probed. Rejects inked but included in bulk shipment.	See Figure 4
2. 100% probed. Electrical and mechanical rejects removed.	See Figure 2 and Figure 3
3. Same as above, but sample tested in a package to meet negotiated acceptance criteria.	See Figure 2 and Figure 3

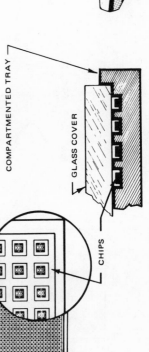

FIGURE 4 – STRAW-PAK PLASTIC VIAL BULK SHIPMENT

The Straw-Pak is a vial encompassing a straw that has one end closed. The chips are inserted in the straw, and then the straw is heat and placed in the plastic vial for shipment.

WAFER OPTIONS

Motorola unencapsulated transistors may be obtained in wafer form. The information in Table II gives the various specification verification and packaging options.

TABLE II – Specification Options

WAFERS	Shipping Options
1. Sample probed. Guaranteed minimum yield.	See Figure 6
2. 100% probed. Rejects inked.	See Figure 6
3. 100% probed. Rejects inked, scribed and broken. Wafer is placed between two sheets of mylar or filter paper and vacuum sealed in a plastic bag.	See Figure 5

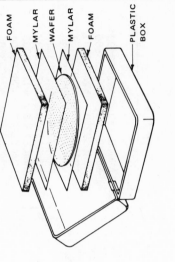

FIGURE 5 – PLASTIC BAG SHIPMENT

Wafer is 100% probed. Rejects inked, scribed, and broken. Wafer is placed between two sheets of mylar and vacuum sealed in a plastic bag.

FIGURE 6 - WAFER SHIPMENT (UNSCRIBED)

FOAM
MYLAR
WAFER
MYLAR
FOAM
PLASTIC BOX

Wafers are shipped between two layers of mylar, sandwiched between two layers of polyfoam pressed together in a plastic box. This prevents movement or damage to the wafer.

 MOTOROLA *Semiconductor Products Inc.*

Figure 4.23. Continued

MOTOROLA

Microcircuit Components

UNENCAPSULATED
SILICON
SWITCHING AND
AMPLIFIER
TRANSISTORS

DECEMBER 1969 — DS 1516

MMCS2192	MMCS3499
MMCS2193	MMCS3500
MMCS2221	MMCS3501
MMCS2221A	MMCS3634
MMCS2222	MMCS3635
MMCS2222A	MMCS3636
MMCS2906	MMCS3637
MMCS2906A	MMCS3903
MMCS2907	MMCS3904
MMCS2907A	MMCS3905
MMCS3250	MMCS3906
MMCS3250A	MMCS4400
MMCS3251	MMCS4401

UNENCAPSULATED
SWITCHING AND AMPLIFIER TRANSISTORS

. . . with passivated Annular* construction that provides high reliability and consistent performance. These chips are identical to the chips used in packaged Motorola transistors with 2N prefixes; i.e., the MMCS2192 chip is used in the Motorola 2N2192 transistor. For more detailed characteristic data, please refer to the equivalent Motorola 2N. . . . data sheet.

- DC Current Gain to 100 Minimum
- Breakdown Voltages to 175 Volts
- f_T to 300 MHz

104

MAXIMUM RATINGS

TYPE	VCEO Volts	VCB Volts	VEB Volts	IC mA	Geometry
NPN					
MMCS2192	40	60	5.0	1000	1
MMCS2193	50	80	8.0	1000	1
MMCS2221	30	60	5.0	800	2
MMCS2221A	40	75	6.0	800	2
MMCS2222	30	60	5.0	800	2
MMCS2222A	40	75	6.0	800	2
MMCS3498	100	100	6.0	500	1
MMCS3499	100	100	6.0	500	1
MMCS3500	150	150	6.0	300	1
MMCS3501	150	150	6.0	300	6
MMCS3903	40	60	6.0	200	6
MMCS3904	40	60	6.0	200	6
MMCS4400	40	60	6.0	600	7
MMCS4401	40	60	6.0	600	7
PNP					
MMCS2906	40	60	5.0	600	3
MMCS2906A	60	60	5.0	600	3
MMCS2907	40	60	5.0	600	3
MMCS2907A	60	60	5.0	600	3
MMCS3250	40	50	5.0	200	5
MMCS3250A	60	60	5.0	200	5
MMCS3251	40	50	5.0	200	5
MMCS3251A	60	60	5.0	200	4
MMCS3634	140	140	5.0	1000	4
MMCS3635	140	140	5.0	1000	4
MMCS3636	175	175	5.0	1000	4
MMCS3637	175	175	5.0	1000	4
MMCS3905	40	40	5.0	200	5
MMCS3906	40	40	5.0	200	5
MMCS4402	40	40	5.0	600	8
MMCS4403	40	40	5.0	600	8

Operating and Storage Junction
Temperature Range −65 to +200°C
*Annular Semiconductors Patented by Motorola Inc.

HANDLING PRECAUTIONS

Although the care and handling of unencapsulated semiconductors often require precautions outside the experience of many equipment manufacturers, Motorola warrants that such devices meet or exceed the published specifications, provided three basic requirements are met in the customer's establishment.

1. Such devices are stored in an environment of no more than 30% relative humidity.

2. Devices are die-and-wire bonded in a noninert atmosphere not exceeding 100°C, or in an inert atmosphere not exceeding 400°C.

3. Processing equipment conforms to the minimum standards of equipment normally employed in semiconductor establishments.

Moreover, Motorola's engineering staff is available for consultation in the event of correlation or processing problems encountered in the use of Motorola semiconductor chips. For assistance of this nature, please contact your nearest Motorola sales representative.

Figure 4.23. Continued

ELECTRICAL CHARACTERISTICS ($T_A = 25°C$)

TYPE	BVCEO Volts min	@ Ic mA	BVCBO Volts min	@ Ic μA	BVEBO Volts min	@ IE μA	ICBO nA max	@ VCB Volts	hFE min/max	@ Ic mA	VCE(sat) Volts max	VBE(sat) Volts max	@ Ic mA
PNP													
MMCS2192	40	25	60	100	5.0	100	10	30	100/300	150	0.35	1.3	150
MMCS2193	50	25	80	100	8.0	100	10	60	40/120	150	0.35	1.3	150
MMCS2221	30	10	60	10	6.0	10	10	50	40/120	150	0.4	1.3	150
MMCS2221A	40	10	75	10	6.0	10	10	60	40/120	150	0.3	1.2	150
MMCS2222	30	10	60	10	5.0	10	10	50	100/300	150	0.4	1.3	150
MMCS2222A	40	10	75	10	6.0	10	10	60	100/300	150	0.3	1.2	150
MMCS3498	100	10	100	10	6.0	10	50	50	40/120	150	0.6	1.4	300
MMCS3499	100	10	100	10	6.0	10	50	50	100/300	150	0.6	1.4	300
MMCS3500	150	10	150	10	6.0	10	50	75	40/120	150	0.4	1.2	150
MMCS3501	150	10	150	10	6.0	10	50	75	100/300	150	0.4	1.2	150
MMCS3903	40	1.0	60	10	6.0	10	50	30	50/150	10	0.2	0.85	10
MMCS3904	40	1.0	60	10	6.0	10	50	30	100/300	10	0.2	0.85	10
MMCS4400	40	1.0	60	100	6.0	100	100	35	50/150	150	0.4	0.95	150
MMCS4401	40	1.0	60	100	6.0	100	100	35	100/300	150	0.4	0.95	150
PNP													
MMCS2906	40	10	60	10	5.0	10	20	50	40/120	150	0.4	1.3	150
MMCS2906A	60	10	60	10	5.0	10	10	50	40/120	150	0.4	1.3	150
MMCS2907	40	10	60	10	5.0	10	20	50	100/300	150	0.4	1.3	150
MMCS2907A	60	10	60	10	5.0	10	10	50	100/300	150	0.4	1.3	150
MMCS3250	40	10	50	10	5.0	10	20	40	50/150	10	0.25	0.9	10
MMCS3250A	60	10	60	10	5.0	10	20	40	50/150	10	0.25	0.9	10
MMCS3251	40	10	50	10	5.0	10	20	40	100/300	10	0.25	0.9	10
MMCS3251A	60	10	60	10	5.0	10	20	40	100/300	10	0.25	0.9	10
MMCS3634	140	10	140	10	5.0	10	100	100	50/150	50	0.5	0.9	50
MMCS3635	140	10	140	10	5.0	10	100	100	100/300	50	0.5	0.9	50
MMCS3636	175	10	175	10	5.0	10	100	100	50/150	50	0.5	0.9	50
MMCS3637	175	10	175	10	5.0	10	100	100	100/300	50	0.5	0.9	50
MMCS3905	40	1.0	40	10	5.0	10	50	30	50/150	10	0.25	0.85	10
MMCS3906	40	1.0	40	10	5.0	10	50	30	100/300	10	0.25	0.85	10
MMCS4402	40	1.0	40	100	5.0	100	100	35	50/100	150	0.4	0.95	150
MMCS4403	40	1.0	40	100	5.0	100	100	35	100/300	150	0.4	0.95	150

AC* PARAMETERS

TYPE	Cob* Ccb* pF max	Cib Cib* pF max	hfe @ Ic min	Ic mA	VCE Volts	f MHz	td,tr ton* ns max	ts,tf toff* ns max	Test Circuit Fig. No.
NPN									
MMCS2192	20	—	2.0	50	10	20	-,85	180,60	1
MMCS2193	20	—	2.0	50	10	20	-,85	180,60	1
MMCS2221	8.0	30	2.5	20	20	100	26*(1)	70*(1)	2
MMCS2221A	8.0	30	2.5	20	20	100	26*(1)	70*(1)	2
MMCS2222	8.0	30	2.5	20	20	100	26*(1)	70*(1)	2
MMCS2222A	8.0	30	2.5	20	20	100	26*(1)	70*(1)	2
MMCS3498	10	80	1.5	20	20	100	20,35	300,80	—
MMCS3499	10	80	1.5	20	20	100	20,35	300,80	—
MMCS3500	8.0	80	1.5	20	20	100	20,35	300,80	—
MMCS3501	8.0	80	1.5	20	20	100	20,35	300,80	—
MMCS3903	4.0	8.0	2.0	10	20	100	40,40	200,60	6
MMCS3904	4.0	8.0	2.5	10	20	100	40,40	240,60	6
MMCS4400	6.5*	30*	1.8	10	10	100	18,25	260,35	8
MMCS4401	6.5*	30*	2.3	20	10	100	18,25	260,35	8
PNP									
MMCS2906	8.0	30	2.0	50	20	100	26*(1)	70*(1)	3
MMCS2906A	8.0	30	2.0	50	20	100	26*(1)	70*(1)	3
MMCS2907	8.0	30	2.0	50	20	100	26*(1)	70*(1)	3
MMCS2907A	8.0	30	2.0	50	20	100	26*(1)	70*(1)	3
MMCS3250	6.0	8.0	2.3	10	20	100	40,40	200,60	4
MMCS3250A	6.0	8.0	2.3	10	20	100	40,40	200,60	4
MMCS3251	6.0	8.0	2.8	10	20	100	40,40	240,60	4
MMCS3251A	6.0	8.0	2.8	10	20	100	40,40	240,60	4
MMCS3634	10	75	1.2	30	30	100	480*	720*	5
MMCS3635	10	75	1.6	30	30	100	480*	720*	5
MMCS3636	10	75	1.2	30	30	100	480*	720*	5
MMCS3637	10	75	1.6	30	30	100	480*	720*	5
MMCS3905	4.5	10	1.6	10	20	100	40,40	240,70	7
MMCS3906	4.5	10	2.0	10	20	100	40,40	250,85	7
MMCS4402	8.5*	30*	1.3	10	10	100	18,25	260,35	9
MMCS4403	8.5*	30*	1.8	20	10	100	18,25	260,35	9

① SMALL-SIGNAL CHARACTERISTICS

$(I_C = 10\,mA, V_{CE} = 10\,V, f = 1.0\,kHz)$

TYPE	hie k ohms min/max	hre x 10^{-4} max	hoe μmhos min/max
NPN			
MMCS3498	0.16/1.2	3.0	8.0/120
MMCS3499	0.2/1.5	4.8	16/240
MMCS3500	0.16/1.2	3.0	8.0/120
MMCS3501	0.2/1.5	4.8	16/240
PNP			
MMCS3634	80/720	3.6	-/240
MMCS3635	165/1400	3.6	-/240
MMCS3636	80/720	3.6	-/240
MMCS3637	165/1400	3.6	-/240

*AC parameter values are as specified in the standard 2N data sheets. (encapsulated devices)

(1) Typical Switching Times

PARAMETER LIMITATIONS AND WARRANTY

Probe limitations allow 100% testing of low level dc parameters only. DC parameters have been selected to insure electrical characteristics to an LTPD of 10 and ac parameters to an LTPD of 20. Visual Inspection is performed to an LTPD of 20. See "Visual Inspection Criteria" in General Information Section.

MOTOROLA Semiconductor Products Inc.

Figure 4.23.   Continued

107

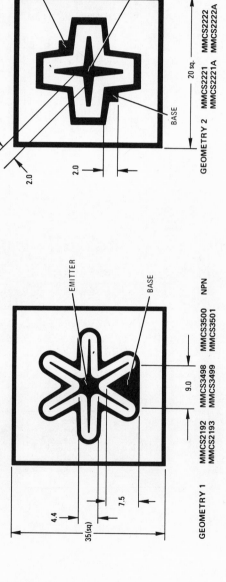

MECHANICAL INFORMATION
MATERIAL – SILICON
FRONT METALIZATION – ALUMINUM
BACK METALIZATION – GOLD (Collector Contact)

ALL DIMENSIONS ARE IN MILS

BASE

EMITTER

BASE

2.0

2.0

20 sq.

GEOMETRY 2

| MMCS2221 | MMCS2222 |
| MMCS2221A | MMCS2222A |

NPN

EMITTER

BASE

9.0

4.4

35 (sq)

7.5

GEOMETRY 1

| MMCS2192 | MMCS3498 | MMCS3500 |
| MMCS2193 | MMCS3499 | MMCS3501 |

NPN

108

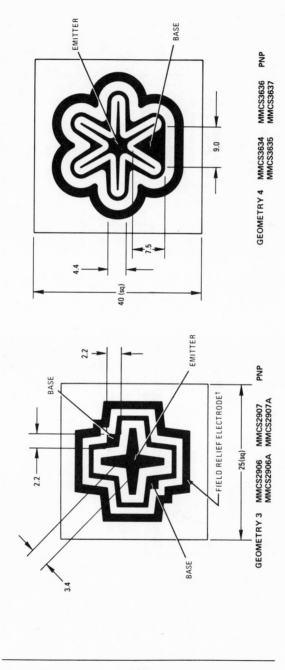

Figure 4.23. Continued

GEOMETRY 6   MMCS3903   MMCS3904   NPN

GEOMETRY 5   MMCS3250   MMCS3251A   PNP
              MMCS3250A   MMCS3905
              MMCS3251   MMCS3906

FIELD RELIEF ELECTRODE†

†Patented by Motorola — Patent No. 3, 302, 076.

(M) *MOTOROLA* **Semiconductor Products Inc.**

110

Figure 4.23.   Continued

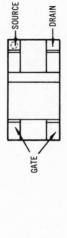

GATE

SOURCE

DRAIN

ENCAPSULANT

BLUE DOT INDICATES THE SOURCE

N–CHANNEL MD1F3069
FIELD-EFFECT MD1F3070
TRANSISTORS MD1F3071

- ● **GENERAL PURPOSE APPLICATIONS**

- ● **ELECTRICALLY SIMILAR TO THE 2N3069, 2N3070, 2N3071**

ABSOLUTE MAXIMUM RATINGS
(at 25°C unless otherwise specified)

Drain-Gate Voltage 50V
Source-Gate Voltage 50V
Gate Current 100mA
Power Dissipation (See Note 1) 400mW
Operating Temperature ...... −65°C to +150°C
Storage Temperature .... −65°C to +150°C

Figure 4.24.   Data sheet—LID packaged, N channel, FET. Courtesy of Dickson Electronics Corp.

ELECTRICAL CHARACTERISTICS (at 25°C unless otherwise specified)

SYMBOL	CHARACTERISTIC		MIN	MAX	UNITS	TEST CONDITIONS
$V_{(BR)DGO}$	Drain to Gate Breakdown Voltage		50		V	$I_D = 1\mu A$, $I_S = 0$
I_{DSS}	Saturation Current Drain to Source	MD1F3069 MD1F3070 MD1F3071	2 0.5 0.1	10 2.5 0.6	mA mA mA	$V_{GS} = -30V$, $V_{DS} = 0$
I_{GSS}	Total Gate Leakage Current	100°C		-5.0 -5.0	nA μA	$V_{GS} = -30V$, $V_{DS} = 0$ $V_{GS} = -30V$, $V_{DS} = 0$
V_P	Pinch-Off Voltage	MD1F3069 MD1F3070 MD1F3071		-10 -5 -2.5	V V V	$V_{DS} = 30V$, $I_D = 1.0nA$
g_{fs}	Transconductance	MD1F3069 MD1F3070 MD1F3071	1000 750 500	2500 2500 2500	μmhos μmhos μmhos	$V_{DS} = 30V$, $V_{GS} = 0$ f = 1kHz
g_{oss}	Small-signal, common-source, output conductance (input shorted)	MD1F3069 MD1F3070 MD1F3071		80 30 7	μmhos μmhos μmhos	$V_{DS} = 30V$, $V_{GS} = 0$ f = 1MHz
C_{gdo}	Gate to Drain Capacitance	MD1F3069 MD1F3070 MD1F3071		2.5 2.5 2.5	pf pf pf	$V_{DG} = -10V$, $I_S = 0$, $V_{DG} = -8V$, f = 1MHz $V_{DG} = -5V$
C_{gso}	Gate to Source Capacitance	MD1F3069 MD1F3070 MD1F3071		5 5 5	pf pf pf	$V_{GS} = -10V$, $I_D = 0$, $V_{GS} = -8V$, f = 1MHz $V_{GS} = -5V$
C_{iss}	Small-signal, common-source, input capacitance (output shorted)	MD1F3069 MD1F3070 MD1F3071		15 15 15	pf pf pf	$V_{DS} = 12V$ $V_{DS} = 8V$, $V_{GS} = 0$ $V_{DS} = 5V$, f = 1MHz
C_{oss}	Small-signal, common-source, output capacitance (input shorted)			1.5	pf	$V_{DS} = 30V$, $V_{GS} = 0$ f = 1MHz
N.F.	Noise Figure	MD1F3069 MD1F3070 MD1F3071		3 3 3	db db db	$V_{DS} = 15V$ $V_{GS} = 0$ $V_{DS} = 10V$ f = 1kHz $V_{DS} = 5V$ $R_G = 1M\Omega$ NBW = 200Hz

NOTE 1. Power dissipation applies when the device is mounted on a 1″ x 1″ x .01″ alumina substrate with 25% metalization in free air at 25°C.

Figure 4.24. Continued

final hybrid. Again, specifying as simple a testing procedure as possible reduces manufacturing costs (which are, of course, passed to the consumer). Stocking spare semiconductor chips involves a small financial investment but allows lower procurement costs through quantity buying. It is always advantageous to consider two vendors for semiconductor requirements. A sole or secondary source should be selected by determining whether the company manufactures the required devices, whether it can supply to a rigid time schedule at reasonable cost, whether the manufacturer will discuss designs using his chips, and whether his test procedure is adequate and flexible. Finally, it is important to know that a given manufacturer's product is reliable and that his firm has the financial stability to stay in business for a long time.

In summary, ready availability and a wide variety of applications are advantages of conventional chips. Chips are made by every major semiconductor manufacturer, usually at attractive prices. Since chips are made on the same production line upon which a vendor produces his discrete components, test and performance data are already available. The disadvantage of conventional chips is that their contacts must be wire bonded in a way that requires expensive equipment and skilled operators, costs money to perform, and is a potential source of system failure. Chips are small, on the order of 8 to 40 mils square by 6 mils thick, and require considerable care in packing, handling, probing, and installation to avoid damage. Additional care must be taken during circuit encapsulation to avoid long-term degradation of the device.

Major Semiconductor Package Configurations. This section outlines the characteristics of and implementation procedures for five major chip semiconductor package configurations.

Conventional Planar Monolithic Chips. Conventional chips used by a manufacturer for incorporation into his own semiconductor packages are made by vacuum deposition processes which create semiconductor regions and overlay gold or aluminum thin film contacts to the top of the chip. The chip is interconnected to the rest of the hybrid circuit by a wire bonding. Additionally, a gold-coated substrate contact is usually brought out the back side of the device; connection to this point is made by bonding the chip to a conducting surface.

The wide availability of conventional chips is their chief advantage. A major disadvantage is the chip and wire bonding procedure (shown in Figure 4.25), which must be done by a skilled operator and is susceptible to both short- and long-term failure.

Flip Chips. Flip chips differ (14–16) from conventional chips in that all of the chip contacts are found on one side of the device. These contacts are heavily metallized solder-coated spheres or pillars about 6 mils in diameter. Flip chips are attached to the substrate by inverting the device so that the contacts touch corresponding lands on the substrate. The lands are then heated, causing solder reflow between the contacts and lands. Thus flip chips do not require fragile wire bonding. The solder reflow operation is performed at a relatively low temperature, whereas three or more high-temperature operations are required by the conventional chip. Flip chips therefore are less expensive to use and result in a higher degree of reliability.

Relatively large contacts on the flip chip allow more complete and automated testing than do contacts on conventional chip. The flip chip is coated with a thick layer of glass, and thus tends to withstand environmental changes better than the conventional chip. The disadvantages of using the flip chip are fewer avaialble types than conventional chips, higher cost, and smaller manufacturers' inventories. Bonding requires careful alignment between chip and substrate. Special indexing marks or alignment fixtures are needed to mount the flip chip correctly. Since flip chips have higher parasitic capacitances, they are less suitable for use at high frequencies than are conventional chips. Two flip chip configurations are shown in Figures 4.26 and 4.27.

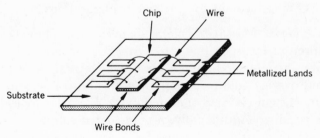

Figure 4.25.   Chip and wire bonding between chip and substrate.

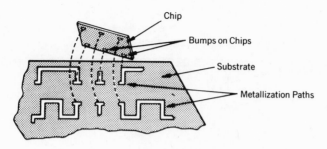

Figure 4.26. Flip chip and substrate—bumps on chip.

Beam Lead Devices. The beam lead device (17–20) of Figure 4.28 is interconnected by relatively large leads or beams, which have been metallized onto the chip by the manufacturer. These leads, made by plating gold, aluminum, platinum, or copper, are usually 4 mils long, 2 mils wide, and 0.2 mils thick and extend over the edge of the chip. An entire slice of several hundred beam lead chips can be fabricated at one time; the chips then can be subdivided by chemical etching from the back of the slice, yielding a slice containing active devices interconnected by beam leads. Thus after etching the entire slice can still be handled as a unit; each chip can be separated from the slice as needed.

Beam lead devices typically are smaller than other devices, containing just the active element and some support for the leads. Since they have large ohmic contacts, a full series of quality control tests can be done on an unmounted beam lead device. Outside connections usually are made by spot welding or ultrasonic bonding. Having very low parasitic capacitance, beam leads are capable of excellent high-frequency performance and can be used in microwave

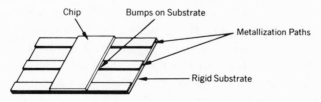

Figure 4.27. Flip chip mounted on substrate—bumps on substrate.

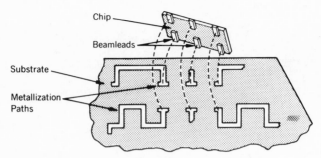

Figure 4.28. Beam lead chip and substrate.

hybrid circuits. They, like the flip chip, are usually coated with a thick, glassy encapsulant. The chief disadvantage of the beam lead device is the relatively small list of available types. However, the beam lead device is becoming increasingly popular in hybrid microelectronics.

Ceramic Flip Chips.   The ceramic flip chip (Figure 4.29), also known as a channel or LID (leadless inverted device), is constructed by mounting a conventional chip within a ceramic block with metallized lands. Wire bonds are made between chip and lands, then the entire ceramic block and encased chip assembly is encapsulated with epoxy or glass frit. A full series of quality control tests is possible, since the ceramic flip chip is very sturdy. The module is mounted to the hybrid substrate by flipping the device and reflow soldering between protrusions on the chip and pads on the substrate.

There is a variation of the ceramic flip chip called the non-inverted channel in which no protrusions are present and all contacts are made to flat metallized areas. The noninverted channel is bonded to the substrate and connections between it and conducting lands are made by a wire bond process.

The disadvantages of ceramic flip chips are size and cost. Ceramic flip chips occupy 10 to 100 times the substrate area of a conventional chip and perhaps 50 to 200 times the volume. Because of the extra connections within the ceramic flip chip, overall reliability of the device is lessened. Production costs make the ceramic flip chip relatively expensive.

Miniature Packages.   In circuits where substrate area is available it is often advantageous to consider using an active device already

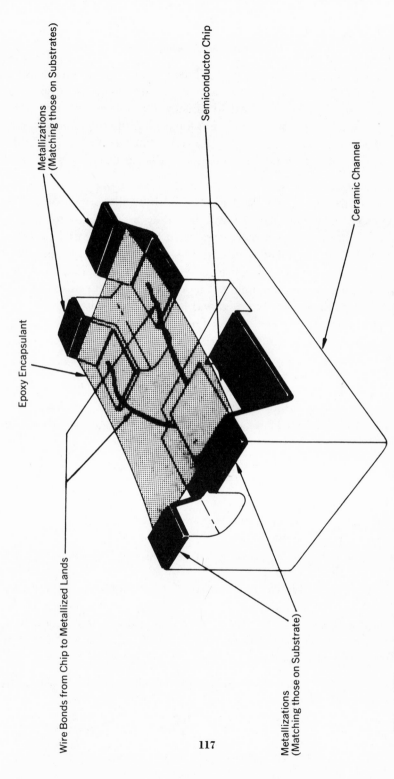

Metallizations
(Matching those on Substrates)

Epoxy Encapsulant

Semiconductor Chip

Ceramic Channel

Wire Bonds from Chip to Metallized Lands

Metallizations
(Matching those on Substrate)

117

Figure 4.29. Channel mounted chip or leadless inverted device. Courtesy of Dickson Electronics, Corp.

mounted in a miniature package. These packages, which measure less than 100 × 100 × 50 mils and usually are plastic encapsulated, can be tested using standard transistor curve tracers and other quality control instruments designed for testing discrete components. Moreover, miniature packages can be attached to the ceramic substrate using conventional transistor soldering techniques.

Summary. Conventional chips are widely available, provide very good high-frequency performance, are physically small, and are inexpensive on a per device basis, but they are difficult to handle and test. The flip chip is more expensive but can be easily handled and tested. The beam lead device is very small, and provides excellent high-frequency performance. More types of beam lead devices are becoming increasingly available. The ceramic flip chip is expensive, readily available, can be handled and tested quite easily, but requires relatively large substrate areas. The miniature package provides approximately the same characteristics as the ceramic flip chip in a slightly larger package.

Bonding Techniques. Many schemes are available for chip attachment and bonding (21–28). The most popular techniques will be discussed in terms of bond strength, costs, production rates, and reliability.

Thermocompression Bonding. Thermocompression bonding is a technique for wiring chips to a substrate. A gold wire is joined to a gold or aluminum substrate contact pad by a process in which heat and pressure are applied to the wire and contact to make a diffusion-bonded cohesive connection. If both metal surfaces are free of contaminants at the interface, high applied heat and pressure will induce a molecular bond.

A tremendously flexible technique for attaching small quantities of chips to a hybrid microcircuit, thermocompression bonding can be accomplished by three basic methods: ball bonding, stitch bonding, or wedge bonding. However, these techniques require a skilled operator to make each interconnection.

Ball Bonding. Thermocompression ball bonding (Figure 4.30) joins a very small diameter (0.5 to 5 mil) gold wire to metallized pads on both chip and substrate. In practice, the wire is fed through a heated tungsten carbide capillary tube and then cut with a hydrogen flame. A meltback occurs on the wire causing a ball to be formed on its end. This ball, while still in a heated state, is forced under pressure to the metallized land. This technique, similar to resistance welding, does not require the passage of an electrical current. Upon completing one ball bond, the capillary tube containing the wire is then moved to the other metallized land, where the wire is again severed with the hydrogen flame, and the resultant ball is forced under pressure to the second metallized land. Thus a ball and stitch bond is made between chip and substrate.

Because of the precision nature of the capillary tip, a relatively high degree of cleanliness is required. The 340°F temperatures required for ball bonding can produce the infamous "purple plague," which sometimes occurs when gold is thermocompression bonded to aluminum. The purple plague is a brittle gold-aluminum compound with poor electrical conductivity that will slowly degrade the bond and cause eventual failure. Purple plague can be avoided completely only by avoiding gold-aluminum bonds. But, by using careful bonding technique and bonding to a large surface area will tend to eliminate the effects of purple plague. A well-made thermocompression bond is reliable and has very high bond strength.

Stitch Bonding. Stitch bonds are also thermocompression bonds. However, in the stitch bonding process of Figure 4.31 the gold wire is not cut by a hydrogen flame. Rather an initial bond is made, then the capillary holder and wire are moved to another metallized land where the wire is not cut by a hydrogen flame but is pressed down by the capillary tube and bonded to a second metallized land. This procedure is then repeated with the wire and capillary holder moved to another metallized land where another bond is made by forcing the lead into the land at high pressure. In none of these intermediate "stitches" is the wire cut, the number of stitches depending on the requirements of the circuit being produced. After the last stitch is made the wire is severed by a hydrogen flame.

Wedge Bonding. In another type of thermocompression bond, the wedge bond, a wedge-shaped tungsten-carbide tool is lowered onto the lead to be bonded, deforming the lead under the tool. This force causes a plastic deformation of the wire as well as cohesion between wire and land. A disadvantage is that wedge bonding tends to weaken and deform the wire.

G T BALL & STITCH OR BALL BOND

Technique Using Gaiser Capillaries . . .

1

BALL FORMED ON WIRE

2

CAPILLARY BRINGS BALL DOWN TO SUBSTRATE TO MAKE BALL BOND

3

CAPILLARY IS RETRACTED, ALLOWING WIRE TO FEED OUT AND MOVED TO SECOND BOND POSITION

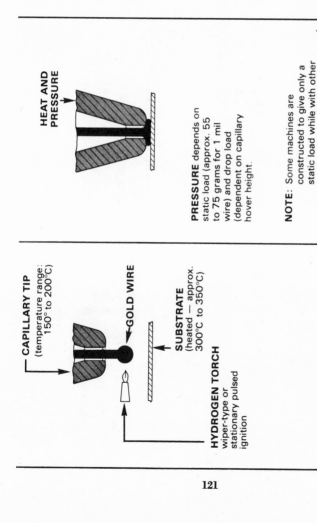

CAPILLARY TIP
(temperature range:
150° to 200°C)

GOLD WIRE

SUBSTRATE
(heated — approx.
300°C to 350°C)

HYDROGEN TORCH
wiper-type or
stationary pulsed
ignition

HEAT AND PRESSURE

PRESSURE depends on static load (approx. 55 to 75 grams for 1 mil wire) and drop load (dependent on capillary hover height.

NOTE: Some machines are constructed to give only a static load while with other machines, one must consider the dynamic load due to the drop from the capillary hover height.

Gaiser suggests a large loop for wire strength. Do not strain wire by stretching tight when feeding out.

Figure 4.30.   Ball bonding.   Courtesy of Gaiser Tool Company.

1. BALL & STITCH OR BALL BOND TECHNIQUE

In this technique, fine gold wire with diameters as small as .0007 inch is fed through the capillary tip. A hydrogen torch forms a ball (approximately 2 to 3 times the diameter of the wire) at the end of the wire. This ball is then brought into contact with the heated substrate and, by the application of controlled heat and pressure, forms a ball bond.

The second bond, a stitch bond, is made by raising the tip, feeding out more wire through the capillary, and moving the tip to the second bond position. The tip is then brought down and heat and pressure are again applied to make the bond. Particular attention should be paid to the capillary tip configuration to achieve the desired stitch bond impression.

4

SECOND BOND

4A. TAIL-LESS

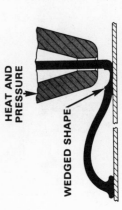

HEAT AND PRESSURE

WEDGED SHAPE

4B. CONTOUR

METHODS USED IN TERMINATING A SEQUENCE OF CONNECTED BONDS:

A HYDROGEN TORCH CUTS WIRE AND FORMS NEW BALL FOR NEXT BOND

To avoid breaking the tail and leaving part of it attached to the bond, use of the Tail-less capillary tip is recommended.

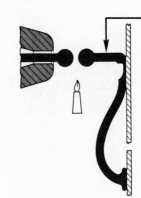

Tail removed by tweezers or automatic tail-puller

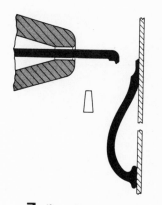

ROUNDED

4C. NAIL-HEAD

SHARP

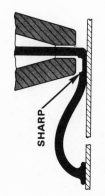

NOTE: For highest bond strength, make two wedge bonds. This assures that bond is not weakened when the tail is removed.

B   TAIL-LESS TERMINATION

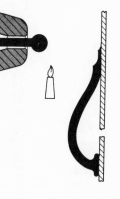

The shape of the Tail-less tip assures easy removal of the tail. This tip design allows for a bonding process and machine design in which no tail is formed. After the second bond is formed, the tip is raised, allowing some wire to feed out.

Then the wire is clamped and the capillary further retracted, breaking the wire at the bond without weakening the bond. Then a new ball is formed.

123

Figure 4.30.   Continued

G MULTIPLE STITCH

T Technique Using Gaiser Capillaries . . .

1

PROCESS BEGINS WITH
WIRE HOOKED UNDER
CAPILLARY, USING
CONTOUR OR NAIL-HEAD
CAPILLARY TIP

(Contour tip recommended
for strongest bond)

2

CAPILLARY BRINGS WIRE
DOWN TO SUBSTRATE TO
MAKE BOND

3

CAPILLARY IS RETRACTED,
ALLOWING WIRE TO
FEED OUT

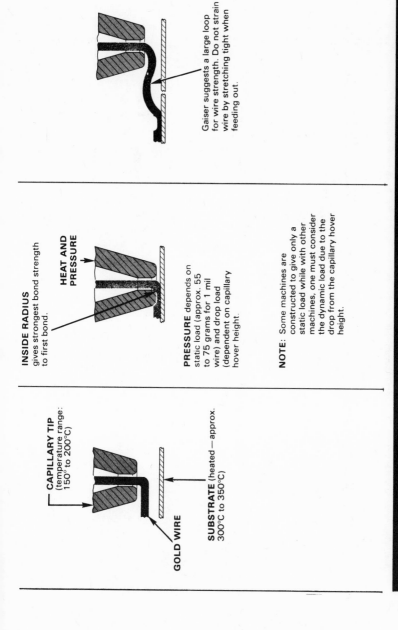

CAPILLARY TIP (temperature range: 150° to 200°C)

GOLD WIRE

SUBSTRATE (heated — approx. 300°C to 350°C)

INSIDE RADIUS gives strongest bond strength to first bond.

HEAT AND PRESSURE

PRESSURE depends on static load (approx. 55 to 75 grams for 1 mil wire) and drop load (dependent on capillary hover height.

NOTE: Some machines are constructed to give only a static load while with other machines, one must consider the dynamic load due to the drop from the capillary hover height.

Gaiser suggests a large loop for wire strength. Do not strain wire by stretching tight when feeding out.

Figure 4.31.   Stitch bonding.   Courtesy of Gaiser Tool Company.

2. MULTIPLE STITCH TECHNIQUE

This technique also features the capillary and the fine gold wire but, in this case, no ball is formed. The process begins with the wire hooked under the capillary, using Contour or Nail-head capillary tip. Capillary brings wire down to substrate, heat and pressure are applied and the bond is made. The next bond is made by raising the tip, feeding out more wire through the tip, and moving the tip to the second bond position. The tip is then brought down and heat and pressure are applied to make the bond. This process can be repeated many times, making a number of stitches, without breaking the wire.

Particular attention should be paid to the capillary tip configuration to achieve the desired stitch bond impression. When a series of stitches has been completed, the wire is broken by subsonic shock, cut mechanically or sheared off (see Figs. A, B and C) leaving the wire hooked under the capillary ready for the next bonding operation.

BREAK

METHODS USED IN TERMINATING A SEQUENCE OF CONNECTED BONDS:

SUBSONIC SHOCK

With tip in position shown below, tip is given a subsonic shock, breaking wire and leaving tail hooked under capillary.

4

SECOND BOND

4A. CONTOUR

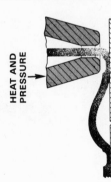

HEAT AND PRESSURE

OUTSIDE RADIUS
gives strongest bond strength
to second bond

4B. NAIL-HEAD

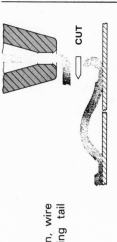

SHARP

NOTE: For highest bond strength, make two wedge bonds. This assures that bond is not weakened when the tail is removed.

NOTE: This process (figs. 1 thru 4) can be repeated many times without breaking wire.

B CUT AND WIPE

With tip in position shown, wire is cut mechanically, wiping tail under capillary.

CUT

C SHEAR

Wire is sheared off on the edge of the package, leaving wire hooked under capillary.

SHEAR

Figure 4.31. Continued

127

All three techniques of thermocompression bonding—ball, stitch, and wedge bonds—are accomplished through the transfer of heat between metallized land and wire with suitable application of force between the work pieces. Heat is applied either through a temperature-controlled heat column beneath the parts undergoing joining or by another heated capillary external to the wire-carrying capillary or by heating this latter capillary tip itself. All techniques using thermocompression bonding require material cleanliness to promote the easy flow of wire through very tiny orifice capillary tubes.

Ultrasonic Bonding. Ultrasonic bonding can be used to mount chips, and attach wires to chips and lands and for other fastening operations. Implemented by forcing a wire against a metallized land and then vibrating the wire at an ultrasonic frequency, usually around 60 kHz, it is useful when heat must be avoided in making the contact or for aluminum to aluminum bonds. Care must be taken to avoid using an excess of ultrasonic energy, which can destroy semiconductor devices. In bonding ultrasonically, a high-frequency scrubbing action microscopically grinds down the roughness between metal surfaces and eliminates any oxides, thus facilitating molecular contact. Many metals, including tin, gold, solder, indium, and silver, may be bonded ultrasonically. The fact that no external heating is required to facilitate this bond is a major advantage over thermocompression bonding. Any heating effects are local and are due to friction. Wire bond strengths in excess of 5 g are possible.

Ultrasonic ball bonding, a process in which a hydrogen flame severs a wire and produces a ball which is ultrasonically scrubbed over a land, is not widely used since an ultraclean work environment is required.

Figure 4.32 shows an ultrasonic wire bonder.

Ultrasonic Spider Bonding. A standard-configuration lead frame can be aligned with a standard pad configuration on a die and all leads can be bonded ultrasonically at the same time or in rapid succession. This technique, although requiring a large expenditure for auto-mated equipment, will produce extensive savings in time and labor. It can also be used in solder-bump flip chip bonding.

Soldering. The main advantages of soldering are the relatively low temperature and simple equipment required to solder bond. Soldering

Figure 4.32. Ultrasonic wire bonding. Courtesy of Weldmatic Division, Unitek Corp.

is the strongest of the bonding techniques, if it is well done. Silver terminations on devices provide optimum solderability and maximum strength of the bond. However, care must be taken to prevent silver leaching into the solder. This is done by using a silver-bearing solder, for instance, a solder containing 62% tin, 36% lead, 2% silver, which melts at 372°F, or a solder of 96% tin, 4% silver, with a melting point of 430°F. Conventional 60–40 solder should not be used.

Reflow soldering, one of the most successful bonding techniques, avoids the use of flux but requires pretinned chips and pretinned substrate pads. In addition to the lead-tin-silver solders, alloys of indium or gallium are used because of their low melting point. However, if the final package is to be hermetically sealed such low-melting-point solders cannot be used.

Welding Techniques. Thermocompression and ultrasonic bonding, both welding techniques, are commonly used in bonding leads in hybrid circuits. Beyond these, parallel gap welding and laser and electron beam techniques can be used to weld chip and leads to the substrate. A discussion of these methods follows.

PARALLEL GAP AND WELDING.   Parallel electrodes are placed under the conductor materials that are to be joined. The electrodes are moved into very close contact with, and under high pressure onto, the bonding area. A current is then passed from the welding electrodes through the conducting materials in such a way as to produce a bond. The electrode materials commonly used for parallel gap welding are copper, tungsten, and molybdenum.

For parallel gap soldering, a pretinned pad and land are joined by solder reflow as current is passed through the joint. The use of flux will result in a uniform, reliable, and more easily made joint. Lead wires between 1 and 5 mils in diameter are commonly bonded by this technique. Many automated methods have been used for parallel gap soldering.

LASER AND ELECTRON BEAM WELDING.   Lasers and electron beams can produce precision microfine joints, but they require relatively expensive equipment. Electron beam welds are performed in a vacuum environment to prevent beam defocusing. Owing to vacuum pumpdown requirements, a high degree of automation is required to make a large number of simultaneous welds, which will insure an economical process. Although laser welds do not require a vacuum environment, inert atmospheres are still advisable to avoid contact oxidation. Both laser and electron beam techniques should enjoy wider acceptance as requirements for smaller microcircuits increase.

Back Bonding. Back bonding attaches the semiconductor chip to the substrate while providing electrical connection to the back side terminal of the chip. Three common types of back bonding are made by reflow soldering, eutectic die bonding, and conductive epoxy adhesive bonding.

Solder reflow must be carefully controlled to avoid air gaps within the bonding layer. Afterward, if the back bond is to be kept intact, the hybrid circuit must not be heated above the solder melting temperature. This technique, however, will allow easy removal of the chip if repair or replacement is needed.

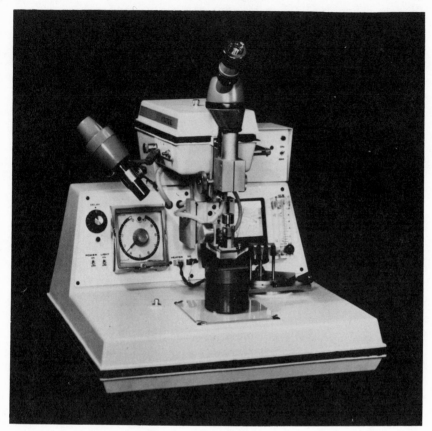

Figure 4.33. Die bonder. Courtesy of Weldmatic Division, Unitek Corp.

Eutectic die bonding (Figures 4.33 and 4.34) is accomplished by holding the chip against the heated substrate until the gold-silicon eutectic alloy on the back side of the chip becomes contact-welded to a gold pad on the ceramic substrate. To be more specific, the substrate is placed on a heated pedestal while the semiconductor chip is scrubbed back and forth to produce cohesion by a physical process not unlike ultrasonic bonding. Since these bonds are made at 380 to 450°C, which are above common solder melting points, relatively high-temperature processes can follow eutectic die bonding.

Conductive plastic epoxy adhesives can also be used for semiconductor chip attachment, as in Figure 4.35. This process is done at curing temperatures of 150 to 250°C. Epoxies that may be pur-

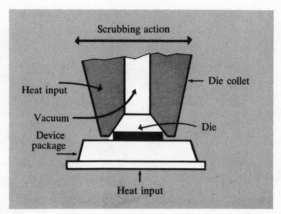

Figure 4.34. Die bonding action. Courtesy of Kulicke and Soffa Industries, Inc.

chased in rigid or flexible forms are easy to use, especially where a crowded substrate leaves no room for the tools of solder reflow or eutectic bonding. Conducting epoxies have the following disadvantages: contact resistance is usually very high; long-term reliability has not been proven; and care must be taken to avoid shorting electrodes when applying conducting epoxy to semiconductor chip

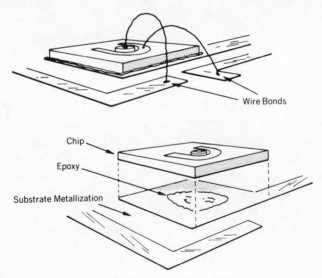

Figure 4.35. Chip and wire transistor mounting with collector bonded with conducting epoxy to substrate. Courtesy of Epoxy Technology, Inc.

and substrate. The use of epoxy is nevertheless a very flexible solution to back bonding.

Flip Chip Bonding. Flip chip bonding, also known as face down bonding, wireless bonding, or registrative bonding, involves physically inverting a chip, placing it in contact with the substrate, and then bonding as in Figure 4.36. Here the conductors on the face of a lid are a mirror image of the conductive lands on the substrate.

As a technique for bonding semiconductor and passive chips to a substrate, flip chip bonding offers two major advantages. First, the cost of assembly is reduced since all bonding is done in one step; and second, the flip chip scheme is inherently more reliable than wire bonding because the total number of interconnections is reduced by a factor of one-half. Unfortunately, flip chip bonded dies do not allow easy inspection of the bonds, since the contacts are eclipsed

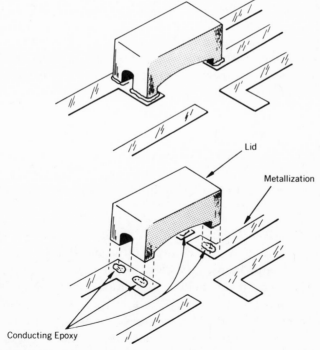

Figure 4.36. Lid bonded in a hybrid circuit using conducting epoxy. Courtesy of Epoxy Technology, Inc.

by the die itself. Air blast and electrical tests have been suggested for inspecting the quality of flip chip bonds.

In design, thermal dissipation must also be considered since flip chip bonding allows a higher density of active devices within a hybrid microcircuit than does the chip and wire technique. A greater amount of heat is generated, and it must be conducted away from the semiconductor chip. High thermal conductivity encapsulants are suggested to facilitate this heat transfer. Flip chip dies are usually glass or epoxy encapsulated to support and protect leads and to prevent corona and voltage breakdown within the microcircuit.

The three most commonly used flip chip attachment schemes, reflow soldering, thermocompression bonding, and ultrasonic welding, are discussed next.

Flip Chip Bonding by Reflow Soldering. The chip or die is bonded by adding solderable material to metallized paths on the die, then inverting, heating, and pressing the die onto a heated substrate. Heating is accomplished by passing chip and substrate through an inert-atmosphere furnace, which heats both substrate and die to a temperature high enough to cause solder reflow. Care must be taken in the reflow soldering technique (as with other flip chip bonding methods) to insure that all chip bumps and metallized substrate lands are coplanar, especially when more than three bumps are to be joined to respective lands.

Flip Chip Bonding by Thermocompression. Both thermocompression and ultrasonic flip chip bonding require very precise alignment of die to substrate. Bonds made by the thermocompression technique are of low cost yet highly reliable. As with thermocompression wire bonding, cohesion is effected by applying heat and pressure to the elements to be joined.

Ultrasonic Flip Chip Bonding. With this technique, the chip is clamped to the substrate and is vibrated ultrasonically, causing a fusing between pads and lands. A variety of material pairs can be ultrasonically bonded; these include aluminum-to-aluminum as well as lead-to-tin, gold-to-tin, silver-to-tin, and cadmium-to-copper. Ultrasonic bonding can produce very strong bonds with tensile strengths of over 40 g per contact. Since such high bond strength does not facilitate die removal for repair, in practice ultrasonic bonding is accomplished at a lower energy level.

Other Flip Chip Bonding Methods.   Many other flip chip configurations have been proposed and are currently in use. The fundamental techniques, however, are reflow soldering, thermocompression, and ultrasonic bonding. The other approaches, variations of these three, are attractive in production line processes where a large number of bonds must be made to identical or similar configurations. These other flip chip configurations include the SLT process, which reflows nickel-plated copper spheres from the chip into substrate areas of lead-tin solder, and the controlled collapse method, which replaces the solid copper ball with a pad of lead-tin solder, producing a reliable, replacable, ductile joint.

A laminated pad technique prevents solder collapse by coating a silver contact with lead-tin solder. Bonding is then made ultrasonically with the solder acting as a bonding agent and the silver section providing structural rigidity. Solder collaspe at the contacts can also be prevented by using a curtain of nitrogen surrounding the periphery of the contact. The cooling effect of the gas prevents solder migration.

The rapid pulse technique in which a controllable pulse of current is passed through the contacts, allowing heating only at the interface between contacts, also prevents solder collapse. Standoffs placed between the solder pads can also be used to provide mechanical support and prevent solder collapse.

Other variations include the use of decals, bonding to an encapsulated lead frame, and several moat techniques. The most common of these is known as "filling the moat." Here the die is inserted into the bottom of a well or "moat," which is then filled with an inert insulating material, and interconnections are made by metallizations between the substrate lands and device pads or by using other techniques such as overlay circuitry or decals. Care must be taken to match the coefficients of expansion of the substrate, die, and filling material. Another production line technique called the STD process forces the chip into a warm thermoplastic material. This technique is similar to "filling the moat" in that contacts can be evaporated between die and substrate. Again thermal expansion must be controlled, and the method does not allow easy replacement of faulty chips.

Another widely used production technique based on beam leading where a large number of leads must be simultaneously bonded is known as "coining and anvils."

REPAIRABILITY AND REPLACEABILITY

Imbedded devices such as those used in the STD process, as well as back bonded dies, are more difficult to replace than are flip chip or beam lead semiconductors. When removed, back bonded devices usually leave residues of gold/silicon or other eutectics.

Wire bonded and spider devices can be replaced on a lead-by-lead basis. Flip chips bonded by reflow soldering can be removed by remelting the solder contacts; flip chips bonded ultrasonically or by thermocompression usually can be sheared off by applying a torque between chip and substrate. Care must be taken to initially produce a joint weaker than the adhesion between land and substrate.

Other possible repairs include fixing broken conductor lines, which can be done by using a fine brush, applying and refiring a paste touch-up, or by the application of preforms and decals. Jumper wires can also be reflow soldered or welded to repair conductor gaps. Localized plating techniques can also be used. If sufficient land area is available, thermocompression bonded leads that prove faulty can be redone.

SUMMARY

The catalog of passive and active chip components was presented in Chapter 4. Chip resistors and their advantages were mentioned, as were microelectronic inductors, high K and high Q capacitors. NPO capacitors were singled out for high-frequency applications and K1200 chips for bypass and coupling functions. Variations in capacity with temperature, applied DC voltage, and frequency were mentioned, as were dissipation factor and insulation resistance. The discussion of semiconductor chips included data sheets, procurement testing, and package configurations. Thermocompression, ball, wedge, stitch, ultrasonic, solder, parallel gap, and laser and electron beam bonding techniques were included, as were flip chip methods. Circuit repairability was also discussed.

REFERENCES

1. M. L. Topfer, "Uses and Advantages of Chip Resistors," *SST*, Vol. 13, No. 11, pp. 53–54, November 1970.
2. U. S. Capacitor Corp., "Application Notes," pp. 1–27.

3. D. W. Hamer, "Ceramic Capacitors for Hybrid Integrated Circuits," *IEEE Spectrum*, Vol. 6, No. 1, pp. 79–84, January 1969.

4. A. L. Houde, "High 'Q' Silicon Monoxide Thin-Film Chip Capacitors for Hybrid Microelectronics," *SST*, Vol. 11, No. 5, pp. 47–55, May 1968.

5. D. E. Maguire, "Tantalum Chip Capacitors," *Proc. 1968 Hybrid Microelectronics Symposium*, pp. 125–135, October 1968.

6. R. A. Lambrecht, "Unencapsulated Solid Tantalum Capacitors for Hybrid Circuit Applications," *1970 International Hybrid Microelectronics Symposium*, pp. 4.6.1–4.6.6, November 1970.

7. S. Slenker, "Micromagnetics in Film Circuitry," *Proc. First Technical Thick Film Symposium*, pp. 157–168, February 1967.

8. J. Abernathy, "Active Devices in Hybrid Integrated Circuits," *IEEE Workshop on Thick Film Hybrid IC Technology*, pp. 4.1–4.14, March 1968.

9. G. R. Broussard, "Supplying Semiconductor Chips to the Microelectronic Industry," *Proc. 1969 Hybrid Microelectronics Symposium*, pp. 1–3, September 1969.

10. L. Stern and I. Carrol, "The Why's and How's of Semiconductor Chips," *SST*, Vol. 13, No. 11, pp. 48–52, November 1970.

11. G. A. Hardway, "Applications of Laser Systems to Microelectronics and Silicon Wafer Dicing," *SST*, Vol. 13, No. 4, pp. 63–67, April 1970.

12. C. S. Stephens, "Getting Started Handling Active Devices," *Proc. IEEE Conv.*, pp. 420, 421, March 1970.

13. B. M. Austin, "Chip Handling Equipment for Hybrid Microcircuits," *SST*, Vol. 13, No. 11, pp. 55–57, November 1970.

14. W. B. Hugle, J. L. Bamber, and D. G. Pedrotti, "Flip Chip Assembly," *SST*, Vol. 12, No. 8, pp. 62–67, 99, August 1969.

15. P. Lin and S. In, "Design Considerations for a Flip Chip Joining Technique," *SST*, Vol. 13, No. 7, pp. 48–64, July 1970.

16. H. J. Shah and J. H. Kelly, "Effect of Dwell Time on Thermal Cycling of the Flip-Chip Joint," *1970 International Hybrid Microelectronics Symposium*, pp. 3.4.1–3.4.6, November 1970.

17. R. K. Field, "The New World of 'Leaded' Chips," *Electronic Engineer*, Vol. 27, No. 8, pp. 100–107, August 1968.

18. L. K. Keys, F. J. Francis, and A. J. Russo, "Bonding Conditions and Tests and Fine Line Techniques for Attaching Beam-Lead Devices to Hybrid Integrated Circuits," *1970 International Hybrid Microelectronics Symposium*, pp. 7.6.1–7.6.10, November 1970.

19. M. P. Eleftherion, "Handling and Bonding of Beam Lead Sealed-Junction Integrated Circuits," *Proc. 1968 Hybrid Microelectronics Symposium*, pp. 323–332, October 1968.

20. M. F. Ramano, "Beam Lead Decision Still Crucial," *Hybrid Microelectronics Review*, pp. 1, 2, July 1970.

21. M. Ohanian, "Bonding Techniques for Microelectronics," *SCP and SST*, Vol. 10, No. 8, pp. 45–52, August 1967.

22. L. F. Miller, "A Survey of Chip Joining Techniques," *SST*, Vol. 12, No. 8, pp. 47–52, 99, August 1969; Vol. 12, No. 9, pp. 33–41, September 1969.

23. L. F. Miller, "A Critique of Chip-Joining Techniques," *SST*, Vol. 13, No. 4, pp. 50–62, April 1970.

24. J. P. Budd, "Die and Wire Bonding Capabilities of Representative Thick Film Conductors," *SST*, Vol. 12, No. 6, pp. 59–63, June 1969.

25. E. F. Koshinz, "Die Bonding Principles and Considerations," *SST*, Vol. 11, No. 8, pp. 47–50, August 1968.

26. E. F. Koshinz, "Face Bonding: What Does it Take?," *SST*, Vol. 12, No. 8, pp. 53–57, August 1969.

27. L. F. Miller, "Joining Semiconductor Devices with Ductile Pads," *Proc. 1968 Hybrid Microelectronics Symposium*, pp. 333–342, October 1968.

28. P. M. Uthe, "Variables Affecting Weld Quality in Ultrasonic Aluminum Wire Bonding," *SST*, Vol. 12, No. 8, pp. 72–77, August 1969.

Chapter 5

DESIGN COMMENTS

Successful reduction of a discrete component breadboard to thick film circuit size requires design skill and knowledge of the hybrid production process. The designer must also be familiar with available thick film pastes and microcomponents before attempting an overall layout of the hybrid.

This chapter presents guidelines for hybrid circuit design with component value tables, conductor layout rules, thick film resistor designs and package selection hints, a comparison of production costs and yields, and other design suggestions.

Hybrid microcircuit synthesis is begun by partitioning an electronic system into digital or analog subfunctions identified by "blackbox" or performance requirements. Considering each subsystem, circuit designers produce an initial schematic which is then constructed as a "breadboard," tested, and usually modified. Breadboarding allows a comparison between blackbox performance specifications and actual performance obtained with a physically realized circuit. The circuit design phase typically uses two to four man-months of engineer and technician time.

Microcircuit layout including conductor, resistor, and dielectric film geometries follows the successful testing of the discrete component breadboard. As layout is proceeding, the designer establishes initial specifications on components and packaging. If a given design is to be produced in quantity, statistical distributions of component values, tolerances, and temperature coefficients, which can be obtained from ink and device vendors, are needed. Computer-aided design programs including ECAP and SCEPTRE can identify worst case design combinations and predict overall yields. This phase generally takes two weeks to one man-month of an experienced engineer's time.

Hybrid prototype development is begun next: masks and screens are made; substrates are screened and fired; chip components are attached; and the resultant prototype hybrid is tested. Changes in microcircuit layout are normal throughout this phase. When satisfactory prototypes are obtained, circuit production is started in-house or by an outside custom hybrid vendor. Figure 5.1 diagrams the thick film prototype process (1).

CIRCUIT LAYOUT

The central step in the process that begins with blackbox specifications and ends with the insertion of a hybrid microcircuit into an electronic system is the layout of a hybrid microcircuit based on a breadboarded discrete component circuit. After both package type and component values have been specified, the designer must decide whether to use one or both sides of a substrate or a multilayer technique. A tradeoff occurs here because cost is more than proportional to circuit complexity. Designing on both sides of the substrate requires extra interconnections and the dissipation of additional heat, but it will significantly reduce module size (2). Factors to be considered during layout include real estate available within each package type, the placement of hot components, provision for thermal compensation tracking of resistors, conductor and land dimensions, and conductor spacing. Resistor and dielectric inks, the physical sizes of screened components, and all passive and active device chips are chosen at this time (3).

The "footprint" of each component is used to estimate total required substrate area. A footprint includes the actual substrate area occupied by the device, pad terminations, space surrounding

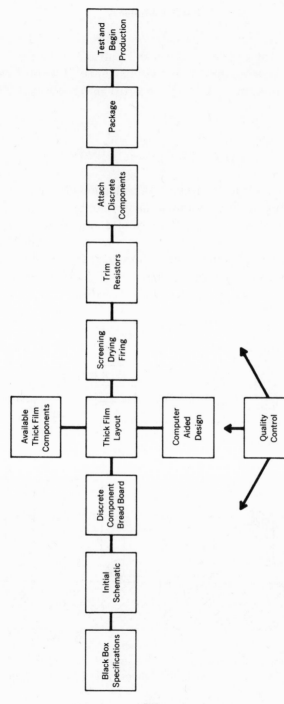

Figure 5.1. Thick film prototype process.

the component, and redundant termination areas necessary for removal and replacement. Figures 5.2 and 5.3 show representative footprints for passive and active components. (Footprint areas for various types of semiconductors are listed in *The DuPont Thick-Film Handbook*.)

CIRCUIT COMPONENTS

Thick film prototyping begins with the selection of substrate material, usually 20-mil-thick, 96% alumina having camber less than 5 mils/in. and surface finish of 25 μin. If thermal conductivity is important, beryllia—at extra cost—is specified. However, it should be remembered that most thick film pastes are designed for alumina substrates.

Resistor pastes are next chosen with odd sheet resistivities blended from two or more pastes. Power dissipation requirements

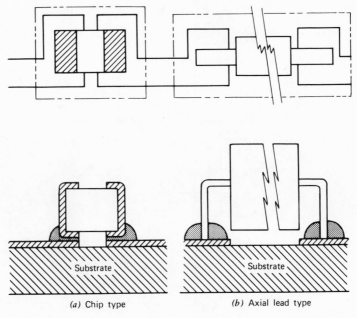

(a) Chip type   (b) Axial lead type

Figure 5.2. Footprints for add-on components such as resistors and capacitors. Courtesy of E. I. DuPont de Nemours and Company (DuPont Thick Film Handbook, p. 23).

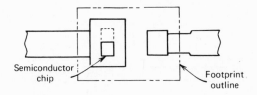

(*a*) Standard chip (face—up) diode.

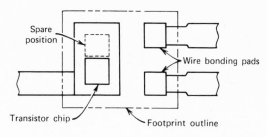

(*b*) Standard chip (face—up) transistor.

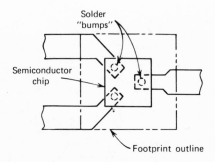

(*c*) Flip chip (face—down) transistor.

Figure 5.3.   Representative semiconductor footprints. Courtesy of E. I. DuPont de Nemours and Company (DuPont Thick Film Handbook, pp. 26, 27).

143

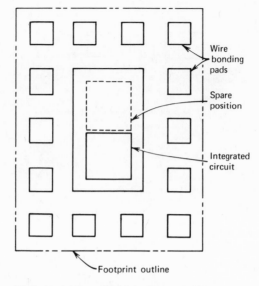

(d) Standard integrated circuit (face–up) chip.

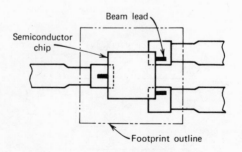

(e) Beam–lead chip transistor and
double diode chip.

Figure 5.3. Continued

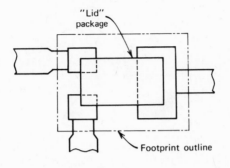

(f) Leadless inverted device.

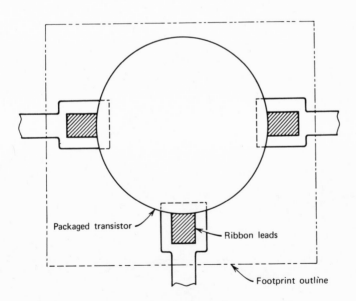

(g) Packaged microtransistor.

Figure 5.3. Continued

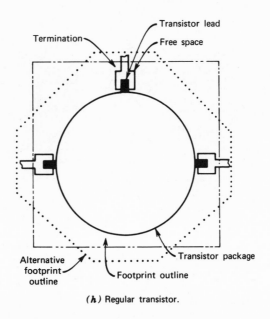

(h) Regular transistor.

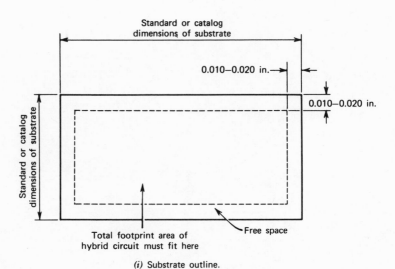

(i) Substrate outline.

Figure 5.3. Continued

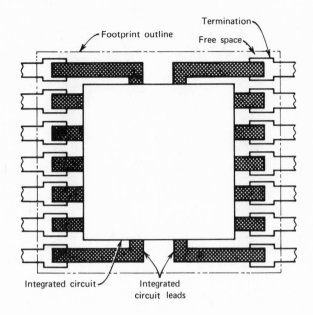

Termination

Footprint outline

Free space

Integrated circuit

Integrated circuit leads

(j) Flat—pack integrated circuit.

Figure 5.3. Continued.

and material sheet resistivity determine the screened size of film resistors, which have resistances from 10 Ω to 10 MΩ with as-fired tolerances of $\pm 10\%$. Trimming can reduce tolerance to better than $\pm 1\%$. TCRs of 100 ppm/°C are usual but can be reduced to ± 25 ppm/°C. Resistor tracking, that is, similarity in change in resistance value between cofired resistors as temperature changes can be held to changes of 50 ppm/°C with tracking to 10 ppm/°C obtainable on a special basis. Power dissipation capabilities are 10 to 40 W/in.². Drift after 1000 hours of operation at 150°C can be held under 1%. Economic reasons dictate avoiding film resistance values below 10 Ω and above 1 MΩ where possible. Most hybrid circuits are fabricated with no more than two resistor screening and firing operations. Screened and fired resistor characteristics are listed in Figure 5.4.

While screened and fired capacitors can be made in capacitances to 15,000 mF and working voltages to 35 V dc, it is doubtful whether a screened and fired effort is worthwhile unless a circuit requires many capacitors. MOS-type chip capacitances of 0.5 to 1000 pF at

Resistance range	1Ω to 10 MΩ
Tolerance	±10%, ±5% special ±0.1% precision trimming
TCR	±100 ppm/°C, ±25 ppm/°C special
Tracking between cofired resistors	10 ppm/°C
Power rating	10 to 50 W/in.^{2}
Maximum voltage	100 V
Operating temperature range	−55 to +150°C
Drift	<1%/1000 hours at 150°C
Noise	−10, −20 dB/decade
Resistor length and width	0.020 × 0.020 in. to 0.040 × 0.040 in.
Resistor spacing	0.040 in.
Resistor-capacitor spacing	0.050 in.

Figure 5.4. Characteristics of typical screened and fired resistors.

tolerances of 10% with working voltages of 40 V and 5 pF to 0.47 mF, 10% tolerance ceramics with working voltages of 50 V are also available. NPO capacitor chips for high-frequency and critical temperature stabilization applications are physically large and thus avoided if possible (4).

Tantalum-slug capacitors from 1000 pF to 250 mF at voltages from 2 to 35 and ±10% tolerances are useful for coupling and bypass functions. Characteristics of capacitors for thick film applications are summarized in Figure 5.5.

Microcircuit "chip" inductors and transformers, typically 75 mils in diameter and 175 mils long with Q's of 7 to 25 and inductances to 5 μH, are occasionally found in thick film microcircuit, as are screened and fired inductors for VHF, UHF, and microwave applications. When possible transformer coupling is replaced by direct RC coupling. Large inductors can be eliminated by active filters or gyrators.

Silicon planar devices including digital and linear IC's are available as dice, chips, or packaged components from almost all semiconductor manufacturers for incorporation into the hybrid.

CONDUCTOR DESIGN RULES

Thick film conductor materials have sheet resistivities of 0.1 to 0.005 Ω/□. A film conductor can be coated with solder to allow solder

Type	Tolerance	Capacity	Voltage Rating	Use
MOS	±10%	1 pF to 0.001 mF	40 V	Signal
Ceramic	±5%	5 pF to 0.001 mF	25 to 100 V	Signal
	±10%	47 pF to 0.47 mF	25 to 100 V	Signal and bypass
Tantalum	±10%, ±20%	1000 pF to 220 mF	35 to 2 V	Bypass
Thick film	±20%, ±10%	10 to 20 pF (NPO)	100 to 250	Signal
screened	special	8000 pF/cm^2 (K1200)		
and fired				

Figure 5.5. Characteristics of chip and screened capacitors for thick film applications.

reflow bonding of leads and components and to reduce conductor resistance. Ordinarily, conductor line widths are 4 to 10 mils, but etched metal masks can produce conductors as narrow as 2 mils. Parallel conductors are usually separated by 8 to 10 mils. In special cases this can be reduced to about 5 mils. Terminal or bonding pads, which must be large enough to insure proper contact for the 5 to 20 mil leads used on attached chips, are commonly 20 $\times$ 30 or 30 $\times$ 40 mils. Contacts for TO5 headers are 10-mil diameter circles. Margins of 5 mils around a pad and spacings of 20 mils between adjacent land areas are minimum specifications for insulation. Conductive lines are kept 10 mils away from the edge of a substrate or hole to avoid damage in handling. Line-width conductors of 20 mils can carry a maximum current of 1 amp. Parasitic capacitances of about 0.5 pF/in. of conductor length are developed when parallel conductors are separated by 10 mils. Interconnection wire lengths. should be less than 100 mils to avoid short circuits caused by shock or vibration. These characteristics are summarized in Figure 5.6.

To achieve good insulation between conductors that cross over each other, dielectric inks should overlap the conductor intersection area by at least 10 to 15 mils on all sides. Since typical conductors have line widths of 5 to 10 mils, crossover dielectric windows should extend to about 20 mils on each side. Crossover inks have approximate dielectric constants of 9 and 1 to 2% dissipation factors. The conductor crossover is depicted in Figure 5.7 (5–7).

RESISTOR DESIGN

Resistor design (8–13) is a compromise between resistor value, number of resistor screenings per substrate, ink costs, resistor size,

Line width	4 to 20 mils
Line spacing	5 to 20 mils
Bonding pad size	20 to 50 mils each side
Maximum current	1 amp for 20-mil-wide line
Parasitic capacitance	0.5 pF/in. length for parallel conductors spaced by 10 mils
Conductor sheet resistance	$<0.010\ \Omega/\square$
Crossover area	20-mils for 2; 10-mil-wide conductors
Crossover layers	2 to 5

Figure 5.6. Typical conductor dimensions.

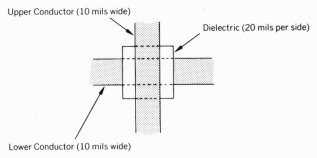

Figure 5.7.   Conductor crossover.

power dissipation, temperature coefficient of resistance, noise, and stability. Resistor design details are shown in Figure 5.8. Commonly selected sheet resistivities of 50 to 30,000 $\Omega/\square$ result in resistances between 5 Ω and 10 MΩ. Careful screening of higher value sheet resistivity inks is difficult and thus should be avoided for double screening operations. Selecting a single ink for all resistors minimizes screening and firing efforts.

To optimize or match the temperature coefficient of resistance between two or more resistors, screening passes should be made in the same direction. Resistor values may be computed by dividing the sheet resistivity in ohms per square by the width of the resistor and then multiplying this figure by the resistor length. Minimum size thick film resistors are 20 mils wide by 20 mils long, with 10-mil lands on end terminations and 5-mil extensions on each side. Some firms specify resistor film overlap of the conductive pattern by 15 mils. The space between any resistor side and adjacent circuitry should be 40 mils or greater to allow trimming.

By screening all resistors to the same design width, process variations will be of the same order of magnitude for each resistor of the same ink and can be compensated for with one change in the screening or firing process. Variations of as-fired resistance values are frequently on the order of $\pm25\%$, although with great care tolerances of $\pm10\%$ can be obtained. For greater precision a means of compensation or trimming is required. The common method is Airbrasion, which alters resistance by removing resistor material. In practice, resistors are screened and fired to lower than design resistance, which is then increased by airbrasion. (Other techniques, mentioned in Chapter 3, trim by lowering resistance.)

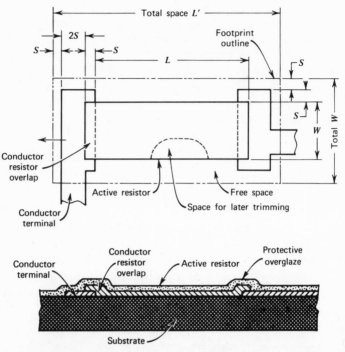

Figure 5.8. Resistor design details. Courtesy of E. I. DuPont de Nemours and Company (DuPont Thick Film Handbook, p. 17).

Airbrasive trimming, which increases the overall length of a resistor relative to its width, can be carried out along a tophat configuration as in Figure 5.9. If resistance calculations are attempted, it should be remembered that the tall portion of the tophat contributes to the conductivity of the resistor much less than do the corners. Since current tends to flow away from the corners and also tends not to flow in the upper portion of the tophat, these regions can be disregarded in conductivity calculations.

Resistor trimming also affects the power dissipation capabilities of resistors. In designing these devices, one can conservatively allocate a power density rating of 10 W/in.² of resistor area. Should resistor stability be of lesser importance (long term ±5% resistance change), a power density figure of 25 W/in.² is allowable. Airbrasion, which increases the power density of the untrimmed segments of a resistor, may mandate increasing the original size of the resistor for adequate post-trim power dissipation.

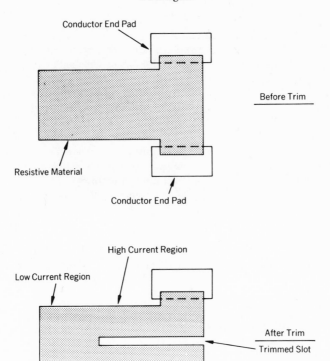

Figure 5.9.   Thick film resistors n tophat configuration before and after trim.

PACKAGES

Standard package configurations tabulated in Figure 5.10 are the TO3, TO5, and TO8 cans and a number of flatpacks. Selection (14–16) of a plastic, metal, or hermetically sealed package depends on environment, circuit requirements, power dissipation, lead configuration, and whether the hybrid has single or multiple substrate layers. The common packages have different power dissipation capabilities. For instance, TO5 cans, 250-mil squares, and 250 × 375 mil, or 375-mil square flatpacks will dissipate 500 mW at 25°C; a 375 × 625 mil flatpack, 750 mW; a 12-lead TO8 can, $1\frac{1}{4}$ W; and a 16-lead TO8 can, or 625-mil square flatpack will dissipate $1\frac{1}{2}$ W.

Package Type	Active Device Complement
TO3	3 transistors or 1 monolithic
TO5	3 transistors or 1 monolthic (per layer)
TO8	
12 lead	8 transistors or 3 monolithics (per layer)
16 lead	15 transistors or 5 monolithics (per layer)
Flatpack	
$\frac{1}{4} \times \frac{1}{4}$ in.	5 transistors or 3 monolithics
$\frac{3}{8} \times \frac{3}{8}$ in.	14 transistors or 4 monolithics
$\frac{5}{8} \times \frac{5}{8}$ in.	22 transistors or 6 monolithics
$\frac{3}{4} \times \frac{3}{4}$ in.	24 transistors or 16 monolithics
1×1 in.	36 transistors or 25 monolithics

Figure 5.10. Thick film hybrid packages.

These figures derate linearly to zero at ambient temperatures of 175°C.

YIELDS AND ECONOMIC FACTORS

Overall yield (17–19) depends largely on the quality of the semiconductor chips as delivered and the degree of control exercized during screening and firing. Thus yield ultimately depends on the skill and competence employed in producing the hybrid.

Thick film screening and firing processes can demonstrate 90% yields for simple circuits and 60 to 80% for more complicated versions. Yields significantly below these indicate that the hybrid production process is out of control.

One of the major elements within the tradeoff between hybrid reliability and cost is the ultimate cost and bother of repairing a defective hybrid versus an initial cost savings made by producing a circuit with higher probability of failure. Certain applications, such as military and space exploration, obviously require higher reliability than does a consumer item. Exact costing of production is difficult, but the cost of factors such as the cost of chip joining and the efficiency of usage of the semiconductor slice are easily determined. (Semiconductors are usually processed in wafer form so the cost per wafer is fixed and the overall cost is dependent on the number of good devices that can be obtained in a single wafer.) Yield affects cost directly, as does the time that facilities are being used in production. Since the time taken in joining substrates and passive components controls cost to a large extent, the time spent at a chip joining station should be minimized.

Although the expense of materials used in fabricating a hybrid is important, for a single circuit the difference in cost between using gold or platinum rather than palladium or silver conductors for example, is quite negligible. Yet considering the production of thousands of units, this difference is no longer trivial. Other economic factors that influence the manufacturing costs include the number of steps in the production process and the degree of required equipment maintenance. The ability to use old but sound tooling and off-the-shelf components can cut costs considerably.

DESIGN SUGGESTIONS

Consideration of the following aspects of design (20–21) should prove useful in producing a minimum-size, low-cost, reliable hybrid:

- Avoid inductors and transformers. Although small chip inductors are available and screened inductors can be fabricated in the microhenry range, the use of *RC* networks for frequency selection or coupling, and the substitution of gyrators or active filters, should be attempted even at the cost of extra stages.

- To keep real estate utilization high, use capacitors of the smallest footprint possible.

- In an *RC* network, resistance should be maximized and capacitance minimized, again to optimize substrate usage. (Capacitance is proportional to area whereas resistance is not.)

- Since additional resistors of the same sheet properties as those already deposited can be added easily, the total number of active devices should be minimized at the expense of extra resistors.

- Choose standard semiconductors that have been proven compatible with hybrid manufacturing techniques.

- Using silicon devices where possible, specify transistor and diode parameters as loosely as can be allowed. Tolerance has considerable effect on yield, particularly for chip and beam lead devices.

- Specifying tight resistor tolerances will reduce yield and increase costs, especially since as-fired tolerances of $\pm 20\%$ can be obtained without a trimming step.
- If possible, specify resistor-ratio tolerances instead of absolute values.
- Specify the tolerance of the overall circuit rather that individual component tolerances.
- Minimize power dissipation while specifying standard package configurations.
- Partition complete circuits for ease of testing, breaking out test points where needed.
- Avoid thin or lengthy chip and wire conductors, which can short circuit during shock or vibration.
- Minimizing the number of long parallel leads will maximize circuit frequency response through minimizing parasitic capacitance.
- Tuneable elements such as potentiometers and variable capacitors should be eliminated if a resistor or capacitor within the circuit can be trimmed to establish a fixed operating condition.
- Using a single sheet resistance ink will minimize production costs and control temperature tracking between resistors.
- Use DC coupling to eliminate large capacitors.
- Do worst-case analysis before finalizing specifications for the hybrid circuit.
- Let the hybrid vendor or in-house production group evaluate the circuit before the design is set. Fully disclose circuit peculiarities uncovered during design and breadboarding phases.

SUMMARY

Chapter 5 has presented guidelines for designing thick film circuits. Included were rules for circuit substrate layout with leads, crossovers, lands, and footprints of passive and active components. Available passives were surveyed, yields were discussed, and a list of design suggestions was presented.

REFERENCES

1. V. K. Gundotera, "In House Thick Film Facility," *Proc. 1970 International Hybrid Microelectronics Symposium*, pp. 7.3.1–7.3.12, November 1970.
2. M. Sobhani and H. R. Isaak, "Design Considerations in Thick Film Hybrid Microcircuits Layout," *SST*, Vol. 12, No. 6, p. 39, June 1969.
3. J. Goldstein, "Hybrid Microcircuit Technology," *IEEE Workshop on Thick Film Technology*, pp. 2–3, 2–4, March 1968.
4. D. T. DeCoursey, "Materials for Thick Film Technology—State of the Art," *SST*, Vol. 11, No. 6, pp. 30–33, June 1968.
5. W. Hayes and J. Van Hise, "How to Design Thick Film Hybrid IC's," *Electronic Engineer*, Vol. 26, No. 8, pp. 72, 73, August 1967.
6. Sobhani and Isaak, *op. cit.*, p. 40.
7. R. E. Thun, "Thick Films or Thin," *IEEE Spectrum*, Vol. 6, No. 10, p. 78, October 1969.
8. P. J. Sanders, "Process Information for the Design of Automatic Resistor Trimming Equipment," *Proc. 1969 Hybrid Microelectronics Symposium*, pp. 197–200, September 1969.
9. Sobhani and Isaak, *op. cit.*, pp. 40–43.
10. DeCoursey, *op. cit.*, pp. 30–31.
11. C. Marcott, "Prototyping Microcircuits," *Electronic Products*, Vol. 10, No. 7, pp. 10–11, December 1967.
12. J. R. Rairden, "Thick and Thin Films for Electronic Applications—Materials and Processes Review," SST, Vol. 13, No. 1, pp. 39–40, January 1970.
13. A. W. Postlethwaite, "Hybrid Thick Film Printed Components—Materials and Processes," *IEEE Workshop on Thick Film Technology*, pp. 5-1–5-9, March 1968.
14. R. C. Musa, "Hybrid Packaging," *IEEE Workshop on Thick Film Technology*, pp. 6-1–6-5, March 1968.
15. R. G. Bristol, "Design Guide to Hybrid Package Size," *Electronic Engineer*, Vol. 28, No. 9, pp. 45, 46, September 1969.
16. Hayes and Van Hise, *op. cit.*, p. 74.
17. J. J. Staller, "Introduction to Thick Film Hybrid Circuits," *IEEE Workshop on Thick Film Technology*, p. 1–8, March 1968.
18. D. Hamer, "Economics of Thick Film Hybrid Microcircuit Production," *Proc. 1970 International Hybrid Microelectronics Symposium*, pp. 6.1.1–6.1.13, November 1970.
19. L. F. Miller, "A Critique of Chip Joining Techniques," *SST*, Vol. 13, No. 4, pp. 51–54, April 1970.
20. Postlewaite, *op. cit.*, p. 5–6.
21. Circuit Technology Inc., "Hybrid Microcircuit Design Manual for the '70's," p. 15.

Chapter 6

HYBRID MICROELECTRONIC CIRCUIT PRODUCTION

In Chapters 1 to 5 we considered some elements of the hybrid production process: substrates, thick film pastes, component trimming, chip devices, and bonding equipment as well as the overall design transformation of a breadboarded circuit into a form suitable for hybrid production. In this chapter we discuss major elements of the thick film process that were noted earlier: screening, including artwork, and screen and mask making; firing of the film-coated substrate; packaging; and quality control.

LAYOUT PATTERNS

From the final schematic of a breadboarded circuit, designers produce thick film layouts, 20 times final circuit size, for each layer of thick film material to be screened and fired (1–5). A master layout is drawn for the overall conductor pattern; this is the first thick film to be placed on the substrate. Other layouts are then drawn for capacitor, crossover, and encapsulating dielectric films, that is, for film resistors and for subsequently printed capacitor plates and crossover conductors. These additional patterns are drawn upon translucent sheets placed over the original conductor layout, resulting in a separate master drawing for each film to be screened. Layout drawings are completed by filling conductor areas with draftsman's black adhesive-backed tape or by cutting Rubylith masters. Rubylith is a sheet sandwich material of transparent clear plastic and transparent red or amber-dyed film held together by an adhesive. The colored layer can be cut in the shape of the conductor pattern and stripped away from the transparent back. Figure 6.1 is a sample Rubylith pattern. Rubylith cutting systems (Figure 6.2) can simplify this procedure in production situations.

The master drawings are then photographically reduced to actual size using a precision reduction camera, screen, and bench as shown in Figure 6.3. The resultant positives are then used to

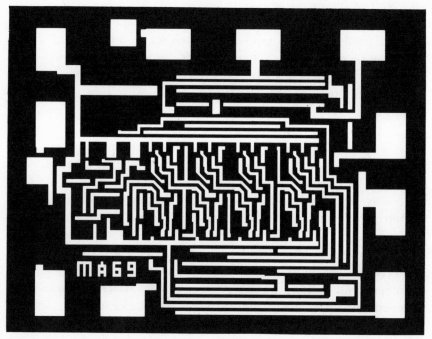

Figure 6.1. A sample Rubylith pattern. Courtesy of Ulano Graphic Arts Supplies, Inc.

produce screens or masks through which thick film layers will be deposited.

SCREENS AND MASKS

Thick film pastes are printed onto the substrate using either a screen or an etched metal mask (6–15), the choice being made on the basis of line width. Screens deliver resolution to 4 mils; masks deliver to 2 mils.

Screens use either direct or indirect emulsion. Direct emulsion is a photographic material applied in liquid form to a 100- to 325-mesh stainless steel screen. The 325-mesh optimizes line definition, whereas the 100-mesh produces best uniformity of deposited paste. A 165-mesh screen is used for dielectrics and 250-mesh is a good compromise for resistive and conductive layers. After application, the emulsion-impregnated screen is dried and then masked with a film positive of one of the layout drawings.

When exposed to a strong carbon arc or halogen light source, exposed portions of the emulsion harden while masked, unexposed areas can be dissolved in water. The resulting screen has open areas through which the thick film is printed. Direct emulsion produces good resolution until the screen material begins to sag, which is usually after 5000 to 250,000 printings.

Another technique, indirect emulsion, will satisfactorily print 100 to 200 substrates. Here emulsion and screen are originally separated. An emulsion/polyester sandwich is first exposed and developed as in the direct method. The wet film is then transferred to an uncoated stainless steel screen. After drying, the plastic backing is pulled away and the indirect/transfer process screen is ready for printing.

With either method, screen mesh is clamped or epoxy bonded to a precision aluminum screen frame or chase (Figure 6.4.) Screen alignment and tension are both important to assure registration of screened and fired films (Figure 6.5 is a photograph of a screen-tension measuring device.)

Photoetched metal mask-deposited films exceed the 4-mil resolution possible with emulsion-coated screens. Two types of

Figure 6.2. Mask being produced by Kongsberg NC mask cutting system, using the Kingmatic 1215 automatic drafting table. Courtesy of Kongberg Systems, Inc.

masks, indirect and direct, are commonly used. The indirect mask is formed by bonding metal foil 1 to 3 mils thick which has been photoetched in the shape of the conductor pattern onto a coarse mesh stainless steel screen. The direct mask is a one-piece molybdenum 0.002-in. foil with a coarse screen or grid pattern etched into the top half of the mask and the conductor pattern etched beneath.

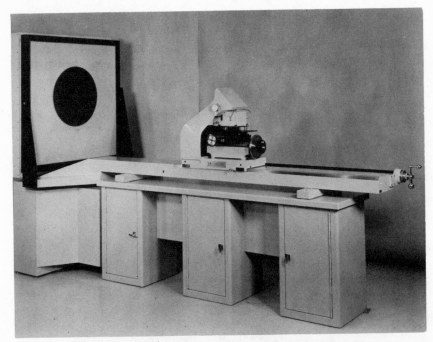

Figure 6.3.   Reduction camera. Courtesy of D. W. Mann Company.

Figure 6.4.   Screen frame jig. Courtesy of deHaart, Inc.

Figure 6.5.   Screen tensiometer. Courtesy of deHaart, Inc.

SCREENING MACHINES

Printing machines are classified as manual, semiautomatic, or auto-matic. The automatic machine performs all screenings and substrate feedings without human attention. A completely automatic machine can decrease operating expense and increase production throughout, but the initial cost of the automatic feed mechanism is very high. Thus most thick film hybrid producers use semiautomatic machines with substrate handling and feeding done by a skilled operator; 600 substrate printing operations per hour can be accomplished this way.

Printing machines contain five basic functional systems of interest: a system to move the substrate into position; a system to hold the substrate rigidly during the printing cycle, usually by applying vacuum to its underside; a screen mounting; an alignment system for adjusting the screen relative to the substrate; and finally a system for supplying ink and moving a squeegee across the substrate. Figure 6.6 is a sketch of a simplified thick film printer.

SCREENING

Paste screening is accomplished either in an off-contact mode or in the contact mode. The off-contact mode with indirect or direct emulsion screens is used more frequently than the contact mode, which requires etched metal masks.

Figure 6.7 diagrams a screen printer showing relative positions of squeegee, ink, screen frame, screen with simplified open mesh conductor pattern, and substrate. Figure 6.8 presents three edge views depicting stages of the screen printing cycle. Figure 6.8*a* shows substrate, paste, squeegee, and the screen with openings at the ready. In Figure 6.8*b* the squeegee is in motion across the screen, deflecting it into contact with the substrate and forcing paste through openings in the screen onto the substrate. After the squeegee passes (Figure 6.8*c*) the screen snaps back, thus shearing the thick film paste which adheres between screening frame and substrate, leaving a printed pattern upon the substrate. This technique is called off-contact, since the screen is held above the substrate at all times except during the deposition of ink. Figures 6.8*b* and 6.8*c* are magnified in

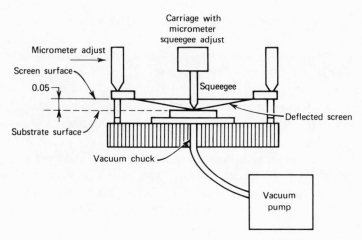

Figure 6.6.   Simplified thick film printer. Courtesy of NASA.

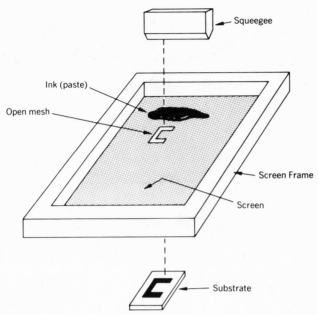

Figure 6.7. Schematic representation of a screen printer. Courtesy of E. I. DuPont de Nemours and Company.

Figure 6.9. Figure 6.10 presents a side view of a thick film printer with squeegee and alignment devices being especially noticeable.

Print thickness, definition, and repeatability depend on substrate uniformity, squeegee speed, squeegee blade angle of attack, and the quality of seal between blade and screen.

INK PREPARATION

The uniformity of final screened and fired films depends greatly on the uniformity of the screened paste. Thus inks should be throughly hand mixed. After initial stirring, ink jars are best stored on a roller mill or orbital mixer which revolves the jar at 2 to 10 revolutions per day, keeping ink particles in suspension and preventing clumping. The better jar rollers pitch and yaw as well as roll the jar.

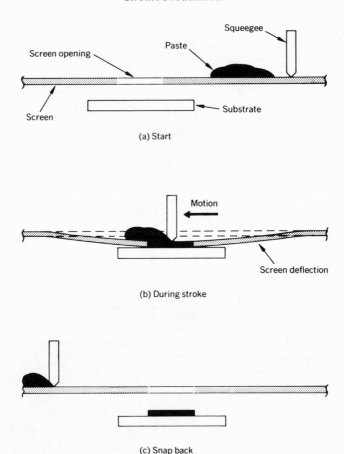

Figure 6.8.   The screen printing cycle. Courtesy of E. I. DuPont de Nemours and Company.

FIRING

After printing, viscous films are allowed to settle in air for about 15 minutes to eliminate the pattern of screen wires in the film. Settling is aided by vibrating the substrate slightly and applying heat in the form of far-infrared energy (longer than 1.5 μ). This curing process removes much of the solvent from the paste, preventing later rapid evaporation of the solvent during firing. The film may be dried in a separate oven or by a series of infrared lamps mounted over the conveyor belt at the entrance to the firing furnace.

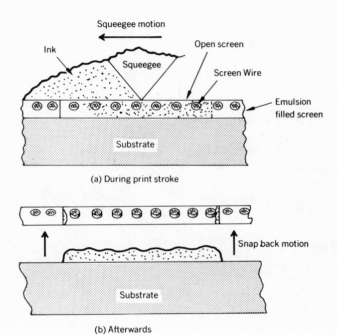

Figure 6.9. The ink transfer process. Courtesy of E. I. DuPont de Nemours and Company.

The substrate then passes into the furnace where it is baked at 700 to 1100°C for 1 to 45 minutes (16–20). Resistors usually are fired at 700 to 850°C, conductors at 850 to 960°C, crossovers at the same temperature, and glazes at 500 to 550°C. This firing cycle burns off the organic vehicle, oxidizes or reduces metallic paste particles, and sinters the glass frit to cause the film to adhere to the substrate. While still in the furnace, the fired substrate is cooled to near room temperature to assure precise and stable resistor parameters.

Screened and settled substrates are fired in a horizontal convetor belt furnace as shown in Figure 6.11. A cross section of the furnace (Figure 6.12*a*) shows the nichrome V or Iconel conveyor belt upon which substrates are carried, Iconel or other metal alloy hearth plate to support the belt and evenly distribute heat, inner wall or muffle of quartz to withstand 1100°C firing temperatures yet not emit gasses that would react with deposited films, and an outer shell

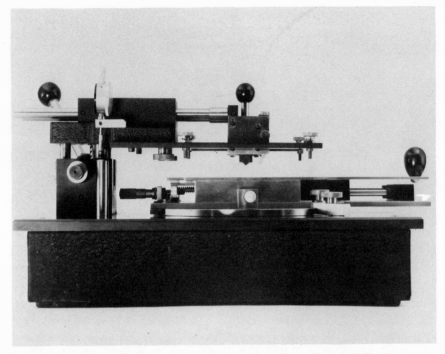

Figure 6.10.   Thick film printer. Courtesy of Engineered Technical Products.

of fire-brick backplates containing electric heating elements. Horizontally (Figure 6.12*b*) the muffle is divided into preheat, high-heat, and cooling zones.

Substrate firing proceeds along a firing profile such as the Watkins 4C-48 furnace shown in Figure 6.13, calibrated in this case to fire 1100°C paste. The profile shown is a plot of temperature vs. distance into the furnace. Since belt speed is constant, Figure 6.13 can be transformed into a curve of temperature versus time the substrate is heated.

Firing begins with a nominal 5-minute 200 to 400°C heating period, which drives off any remaining organic thinners from the paste. Temperature rise in this phase should in no case exceed 200°C/minute to avoid paste bubbling and resultant change of properties. Vapors are carried off by a forced air system and are prevented from entering high heat regions by a baffle door.

The dried film is then sintered by exposure to high heat (1100°C

Figure 6.11. Conveyor furnace. Courtesy of Watkins-Johnson Company.

in the case of the profile of Figure 6.13) for about 15 minutes. This is known as the "plateau" of the profile. Here oxidation/reduction reactions occur until the glass frit melts and seals the metallic particles that constitute conductor and resistor films from further interactions with the atmosphere. The melted glass frit also wets the substrate anchoring the film pattern.

The substrate then is cooled to room temperature and passes out of the furnace for further screening operations, resistor trimming, or the attachment of chip components.

PROCESS CONTROLS

To assure accurate film parameters precise control of temperature, belt speed, atmosphere, and humidity must be maintained throughout the firing cycle.

Temperature gradients across the belt are held to $\pm 1°C$ by using a dense (3 lb/ft^2) nickel-chrome-iron (Inconel) mesh belt.

This belt has high heat capacity, thus the presence of 0.025-in. substrates has little effect on its temperature. Horizontal temperature gradients greater than $\pm°1C$ can cause resistance deviations of 5% or more.

As the substrate passes through the furnace vertically, far-infrared radiation reaches it directly, causing rapid heating. Thermocouple-SCR controls allow temperatures to be set with accuracies of ±0.5 to $\pm10°C$, depending on the furnace.

The time that a film is exposed to a given temperature depends on belt speed. Thus variations in belt speed change the temperature profile versus time for a given substrate. Over the plateau region of profile, belt speed variations of $\pm5\%$ will cause changes of ±15 seconds within, for example, a 5-minute soak time. The effect is even more noticeable in regions where temperature profile is changing rapidly. Controls are used to insure belt speed accuracies to better than $\pm2\%$.

Whereas some processes use air as the firing medium, others require varying percentages of filtered oxygen and nitrogen. Certain furnaces provide facilities for storing, mixing, introducing, and venting these gasses. Since palladium oxide resistor films have been

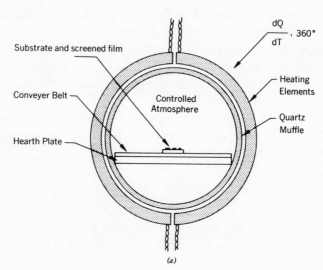

(a)

Figure 6.12(a) Edgeview of a thick film furnace. Courtesy of Watkins-Johnson Company. (b) Horizontal view of a thick film furnace. Courtesy of C. I. Hayes, Inc.

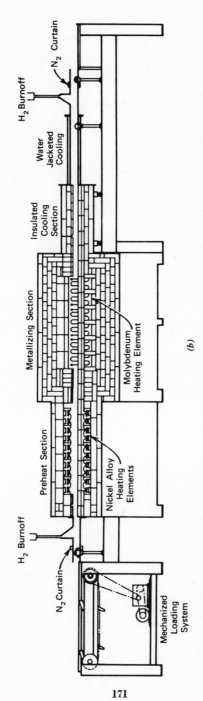

Figure 6.12(a) Edgeview of a thick film furnace. Courtesy of Watkins-Johnson Company. (b) Horizontal view of a thick film furnace. Courtesy of C. I. Hayes, Inc.

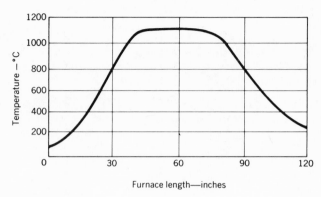

Figure 6.13.   A thick film firing profile. Courtesy of Watkins-Johnson Company.

shown to display a 6% increase in resistance with change in relative humidity from 0 to 60%, a means of controlling humidity in critical firing operations is also useful.

Finally, overall screening and firing yield is a function of the control of firing parameters but also the quality of the ink preparation and screening operation. If yields drop below 90%, it suggested that all elements in the screening and firing cycle be checked to ascertain the cause of variations.

PACKAGE FUNCTIONS

After active and passive components are attached onto the screened and fired substrate, the resultant microcircuit is ready for mounting within a package (21–33). The package protects the circuit from its environment, mechanically supports the contained electronics, provides a heat transfer path from the circuit substrate to the air, and allows external leads to be attached to relatively strong package leads or pins rather than to fragile substrate lands.

Packages are usually tested to determine how well they fulfill these functions. The package seal is tested to measure leakage, moisture resistance, elevated temperature storage, and operation; resistance to chemicals such as fluxes, solvents, and gasoline and oil for automotive applications are also tested. Radiation resistance is checked in radiation-hardened applications. Tightness of seal is determined by the vulnerability of the encased microcircuit components; for example, a tighter seal must be demanded for a substrate containing uncased transistor chips than is needed for a circuit of glazed film resistors and hermetically sealed chips.

Other tests are made of mechanical shock and vibration, thermal shock and temperature cycling, lead strength by twist and pull tests, lead bondability, and the ability to withstand high voltages.

PACKAGE CONFIGURATIONS

Basic packages are the TO5, TO8, and TO18 cans, butterfly, dihedral, platform, and coldweld cases, custom flatpacks with Kovar or ceramic-to-glass seals, and also plastic types. TO3 cans are occasionally used.

TO5 Can. Headers for TO5 and TO8 cans are seen in Figure 6.14. TO5 cans are 350 mils in diameter with 6, 8, 10, or 12 leads possible. This package will hold one or a stack of 180-mil diameter substrates each carrying two to four transistors or one IC chip. The TO5 will dissipate 280 mW at 125°C in free air. TO18 cans are similar to but slightly larger than the TO5 size.

TO8 Can. This package is similar to the TO5 but is approximately twice its diameter. TO8 cans have 12 and 16 lead configuraions and will hold one or a stack of 230- or 300-mil diameter substrates. Depending on size, these substrates will carry six to sixteen transistors or two to five IC chips. The TO8 package will dissipate 1.0 to 1.2 W at 125°C.

Flatpack Cases. These packages hold substrates as small as a 150-mil square or as large as 1.250 × 1.350 in., dissipating 250 mW to 2.5 W, respectively, in free air at 125°C. Smaller flatpacks hold several transistors or monolithics; larger ones can encase over 30 active devices.

Figure 6.15 presents the components of a vertical sidewall case, the header and elevated and flat covers for the package. This case, which can be hermetically sealed, will accommodate substrates from $\frac{3}{4} \times \frac{3}{4}$ in. to 1 × 1 in. Outside dimensions are 0.936 × 1.140 in.

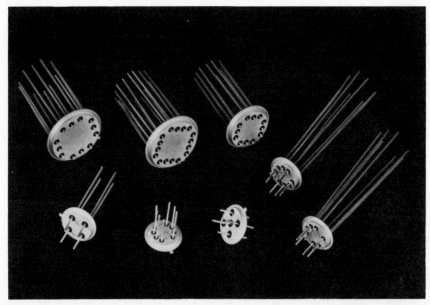

Figure 6.14.   TO5 and TO8 headers. Courtesy of Tekform Products, a Bliss and Laughlin Industry.

Figure 6.15.   Vertical sidewall cases. Courtesy of Tekform Products, a Bliss and Laughlin Industry.

and 1.335 × 1.372 in., respectively. Height available for components (including substrate thickness) is about 115 mils with flat cover or 225 mils with raised cover.

The platform case, as in Figure 6.16, can hold large substrates, up to 1.250 × 1.350 in. Again the case can be hermetically sealed and is available with flat or raised cover; the choice is made by considering circuit height and allowable external thickness. This case can be obtained with as many as 44 lead pins.

Butterfly cases, as in Figure 6.17, may contain substrates from 500 × 500 mils to 875 × 875 mils. Leads are brought out of the sides of this extremely thin package. The butterfly is extremely rugged and can meet all requirements of MIL-STD-883. Figure 6.18 presents a 64-lead butterfly case.

Kovar and glass hermetically sealed flat packs (Figure 6.19), which are extremely thin, can hold substrates from 120 × 160 mils or 150 mils square to 240 × 290 mils. As many as 40 leads may be connected.

Figure 6.16.   Platform cases. Courtesy of Tekform Products, a Bliss and Laughlin Industry.

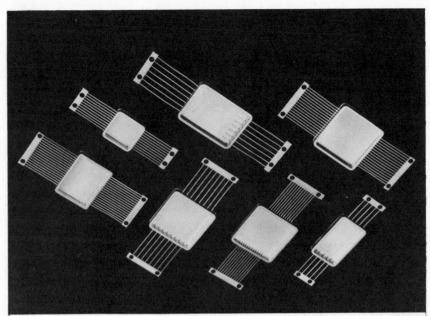

Figure 6.17.   Butterfly cases. Courtesy of Tekform Products, a Bliss and Laughlin Industry.

Figure 6.18.   A 64-lead butterfly case. Courtesy of Tekform Products, a Bliss and Laughlin Industry.

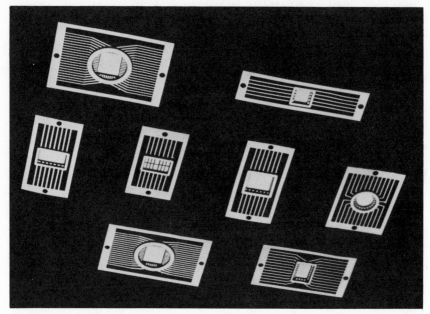

Figure 6.19.   Flatpacks. Courtesy of Tekform Products, a Bliss and Laughlin Industry.

HERMETIC SEALING

Unencapsulated semiconductor chips, which are sensitive to contaminants such as water vapor, hydrogen, oxygen, and sodium ions, require isolation in a hermetically sealed package to avoid circuit performance deterioration. Contaminants can reduce collector-base breakdown voltage, increase leakage current and parasitic capacitance, and reduce current gain in transistors while attacking chip metallizations and chip and wire interconnections.

For hermetic sealing, the package and contents are cleaned thoroughly. Then a dry, inert atmosphere free of contaminants is sealed into the package. Outgassing of vapors from the hybrid or case is discouraged by performing the sealing operation at low temperature.

The hermeticity of the seal is measured according to MIL-STD-883, method 1014 for fine leaks and 1014C for gross leaks. Gross leaks are measured by immersing the package in a 125°C fluid and scrutinizing for bubbles. Fine leaks are detected by attempting to force helium or a radioactive gas tracer into the package under pressure, then measuring the rate of outgassing from the package when it is placed in a vacuum chamber.

The pressure and vacuum stresses of the test are usually much more severe than those that the package will see in service. Since they can destroy seal, package, and encased circuit as well, the question of using such tests must be considered carefully.

PACKAGE SEALING

The method employed to seal a metal package depends on whether the package is to be repairable, the type of package to be sealed, and the degree of hermeticity required. For instance, if a circuit is inexpensively produced such that repairs will consist of discarding the faulty module and circuit, replacing it with an all new hybrid, then the welded enclosure is a practical choice. Repair of a welded enclosure consists of the following expensive operations: grinding off the lid, separating all wire and back bonds between case and substrate, repairing and retesting the circuit outside the package, reinserting the circuit, resealing the package, and testing the repaired and packaged circuit. If the circuit is discardable, welding, which is fast and capable of high yields, is an excellent sealing method.

Resistance welding is excellent but presumes dimensional preciseness of cover and base. Care must also be taken to avoid excessive heating of the enclosed circuit. A variation of this technique, seam welding, concentrates heating to just one point at a time along the interface of cover and base and thus avoids heating problems.

Certain package configurations can be cold welded, a technique that applies high compression to effect cohesion between cover and base.

Circuits costing over $100 are usually packaged with repair in mind. Epoxy or solder seals are used, but both require caution to avoid heating the substrate above the melting point of any solder-bonded substrate connections.

New epoxy compounds no longer outgas as readily as did earlier epoxy-based sealants whose outgassing caused circuit degradation, seal embrittlement, and reactions with nearby metals.

Solder sealing is the most common method of fastening covers to bases in repairable packages. Hand soldering, although slow, unreliable, and messy, is used by almost all hybrid houses since it is inexpensive, flexible in application, and an immediate means of sealing.

Tekform, a major producer of packages, has found three major items to be instrumental to good solder sealing: the quality of gold or tin plating, dimensional stability and fit of case to cover, and the cleanliness of materials to be joined. Figure 6.20 indicates one way of solder sealing a case. A loop or preform of solder placed around the package perimeter is melted by moving an iron rapidly around the edge of the case. Extra solder is introduced as needed to completely fill the cavity between the lip of the cover and the case sidewall.

Figure 6.21 presents edge views of dihedral, coldweld, flatpack, butterfly, and vertical sidewall cases.

PLASTIC PACKAGES

Two approaches are used to provide plastic packaging—casting and transfer molding. In the casting process the hybrid circuit to be encapsulated is placed in a mold and an epoxy resin and catalyst mixture is poured to encase the circuit. This material is then cured in a heated oven until it is hard; the mold halves then are separated and the epoxy-encapsulated hybrid is removed. The casting technique is most useful for low-volume production, since initial equipment cost is low: molds, resins, and oven and resin metering equipment

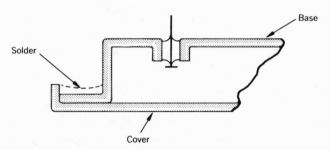

Figure 6.20. Vertical sidewall case solder-sealed. Courtesy of Tekform Products, a Bliss and Laughlin Industry.

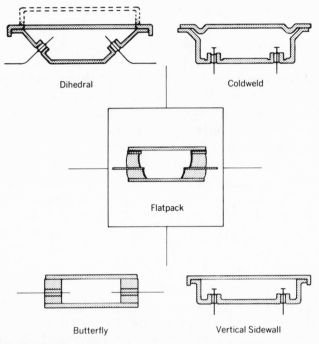

Figure 6.21.   Edge views of common cases. Courtesy of Tekform Products, a Bliss and Laughlin Industry.

are the only elements required. Moreover, the casting technique is flexible enough to accommodate changes in circuit shape, and thus it is useful throughout prototyping. Disadvantages are the cure time and mold cleanliness required.

Transfer molding, the second plastic packaging technique, flows thermosetting plastic in a quasi-liquid state into a mold containing the hybrid and a lead frame. Although it is more expensive initially due to expenditures for a mold and press, the transfer technique lends itself readily to mass production.

The primary advantage of plastic packaging is cost. However, conflicting opinions exist on the quality of plastic packaging. Proponents claim low cost and reliability at least as high as hermetically sealed cans in meeting the applicable specification MIL-STD-750A. Many manufacturers report millions of socket hours without failure even in high-humidity environments. Detractors claim poor heat transfer, nonhermeticity of the seal and failure of fine gold wires when contacted by epoxy potting compounds, or during thermal shock.

Anderson et al. (30) take a middle position, pointing out the dangers of generalizing, especially when thousands of expoy, silicon, and polymer formulations are used as encapsulants. They propose judging the quality of a given package relative to one of three sets of standards: benign military environment, severe military environment, and space environment.

Additionally, this group identifies the following potential causes of failure: high rate of moisture transmission (liquid or vapor); outgassing of the plastic when heated; metallic ion impurities, including Na, K, Li, and others in the plastic; nonstoichiometry of two component plastics; lack of adhesion between package and leads; and stresses caused by plastic shrinkage or by differences in expansion coefficients between circuit components and the encapsulant.

QUALITY CONTROL

Product assurance tests (34–37) are made throughout the hybrid production process: materials and components are tested before screening, firing and assembly; partially completed substrates are tested; and completed packages are checked for their ability to deliver specified performance levels throughout extremes in temperature, atmosphere, and vibration. Tests are instituted to determine probabilities of both short- and long-term failure.

Materials tests are begun by checking substrates for flatness, that is, lack of camber and surface roughness, as well as for conformity with linear specifications.

Screens, emulsions, and squeegee blades are checked periodically for ability to deliver promised pattern resolution. Frame alignment and screen sag are also watched.

Thick film inks are checked for correct viscosity and are kept in suspension by a rolling mill after initial hand stirring. Resistivity, conductivity, and dielectric properties of inks can be checked by performing basic electrical measurements on samples of specially screened and fired test patterns.

Temperature, humidity, and atmospheric content of the firing process can be adjusted by zones throughout the furnace. Belt speed precision and tension also must be maintained.

Tolerance tests can be made on thick film components such as transistors and active devices, as well as on passives such as chip capacitors, resistors, and inductors.

Bond strengths may be checked optically, electrically, or by pull tests. The hermeticity and strength of the package are tested by the methods noted in the preceding section.

After the hybrid is built and packaged, electrical measurements are made of original black box specifications. Unique test programs are designed for each type of hybrid, but they fall into two general categories—digital and linear.

Digital hybrids including RTL, TTL, DTL, HNIL, ECL, and complementary MOS can be tested for tolerance of input and output load currents and voltages—logic 1 and 0, logical functions under specified load conditions and current, shorts between terminals, and for various circuit and clock speeds. Digital test rigs, sometimes automated, contain adjustable high-regulation power supplies, controllable input levels for 0 and 1, output comparitors with selectable 0 and 1 voltage levels, variable clock, data, and phase rates, leakage limit checks between all device pins, and means of comparing device output with a similar device designated as a standard.

Linear hybrids such as the operational amplifier can be tested for input offset voltage and current, bias current, open loop gain, power supply rejection ratio, common mode rejection ratio, input voltage range before saturation, maximum output voltage, and quiescent power supply currents. Similar test schedules can also be made for other linear hybrids.

Both digital and linear circuits are tested throughout a wide environmental range such as the MIL-STD-750A environments listed in Figure 6.22.

SUMMARY

Building upon information gained about substrates, paste characteristics, chip components, and layout presented in previous chapters, the key elements of the thick film production process—artwork, screen making, screening, firing, packaging, and quality control— were discussed. Indirect and direct emulsion screens were presented

Typical MIL-STD-750A Environments

Test	Method	Conditions
Thermal shock (glass strain)	1056	0 to + 100°C
Temperature cycling	1051	−65 to +150°C
Shock	2016	1500 g; 0.5 msec
Variable frequency vibration	2056	20 g 100 Hz −3 k Hz
Random vibration	—	50 g RMS
Vibration fatigue	2046	20 g; 96 hr
Constant acceleration	2006	To 20,000 g
Moisture resistance	1021	
Salt atmosphere	1041	
Terminal strength	2036	
Barometric pressure	1001	2.4 × 10⁻⁶ mm Hg operating
Operating life	1026	1000 hr min. at max. Pd
High temperature storage	1031	1000 hr min. at 150°C
Seal	112	10⁻⁸ cm³/sec

Special Application Environments
Shock—10,000 g, 10 msec Controlled Laboratory Test
 20,000 g, 11 msec Ballistics Test
Vibration—60 g 100-10,000 Hz
Acceleration—10,000 g Attached to Turbine blade

Figure 6.22. Typical MIL-STD-750A environments. Courtesy of Electronic Components Division of United Aircraft.

as standard for screening with fine detail work being done by direct or indirect photoetched metal masks. Firing consists of a settling phase with light heating, a baking phase which applies 700 to 1100°C heat to anchor and encapsulate resistor, conductor, and dielectric films, and a cooling phase to deliver stable film properties. Packages may be hermetically or nonhermetically sealed. The most frequently used packages include various flatpack cases and several TO can varieties. The quality of plastic packages, which can be made by casting for low production requirements or by transfer molding, is still being debated. Product assurance tests were outlined for digital and analog hybrids.

REFERENCES

1. J. M. Conley, "How to Prepare Low Cost Artwork for Thick Film Hybrid Integrated Circuits," *Proc. IEEE Conv.*, pp. 416–417, March 1970.

2. L. Jacobson, "How to Prototype Hybrid Circuit Patterns and Screens at Budget Prices," *IEEE Spectrum*, Vol. 6, No. 7, pp. 82–85, July 1969.
3. B. J. Miller, Jr., "Generating Precision Artwork for I. C. Mask Sets," *Photo/Chemical Fabrication*, August 1969.
4. J. B. Tong, "Mask Manufacture for Integrated Circuits," *SST*, Vol. 11, No. 7, pp. 19–26, July 1968.
5. R. Quick, "A Simple Computer Aided System for the Production of Hybrid Interconnection Artwork and Integrated Circuit Masks," *Proc. 1968 Hybrid Microelectronics Symposium*, pp. 71–85, October 1968.
6. D. R. Kobs and D. R. Voigt, "Parametric Dependences in Thick Film Screening," *1970 Hybrid Microelectronics Symposium*, pp. 5.5.1–5.5.10, November 1970.
7. B. M. Austion, "Thick Film Screen Printing," *SST*, Vol. 12, No. 6, pp. 53–58, June 1969.
8. L. H. Coronis, "Indirect and Direct Etched Metal Masks for Deposition Control and Fine Line Printing," *Proc. 1969 Hybrid Microelectronics Symposium*, pp. 243–251, September 1969.
9. A. V. Ottaviano, "Repeatability in Screen Printing Hybrid Microcircuits," *Proc. 1969 Hybrid Microelectronics Symposium*, pp. 253–262, September 1969.
10. L. H. Coronis, "A Storable Emulsion Screen for Thickness Control and Immediate Response," *1970 Hybrid Microelectronics Symposium*, pp. 5.6.1 to 5.6.7, November 1970.
11. R. E. Trease and R. L. Dietz, "Paste Rheology Can Improve Your Fine Line Printing," *1970 Hybrid Microelectronics Symposium*, pp. 8.4.1–8.4.8, November 1970.
12. J. A. Van Hise, "Process Variable in Thick Film Resistor Fabrication," *Proc. 1969 Hybrid Microelectronics Symposium*, pp. 185–195, September 1969.
13. D. C. Hughes, "Recent Developments in Screen Printing Technology," *NEPCON West*, pp. 1–9, February 1969.
14. R. Ilgenfritz, "Thick Film Hybrid Microelectronic Circuit Technology," *SCP and SST*, Vol. 9, No. 6, pp. 37, 38, June 1966.
15. D. C. Hughes, "The Screen Process Printing Machine and How It's Specified," excerpted from D. C. Hughes, *Screen Printing of Microcircuits* by Precision Systems Company, Somerville, N. J.
16. D. J. Spigarelli, "Selection of Furnace Equipment for Thick Film Firing," *Proc. First Technical Thick Film Symposium*, pp. 37–43, February 1967.
17. M. K. Durham, "Conveyor Furnaces for Microelectronic Production," Watkins-Johnson Company, Bulletin 100265, August 1968.
18. J. H. Beck, "Firing Thick Film Integrated Circuits," *SCP and SST*, Vol. 10, No. 6, pp. 29–32, June 1967.
19. H. H. Nester and T. E. Salzer, "Thick Film Production Techniques," *Proc. First Technical Thick-Film Symposium*, pp. 3–4, February 1967.
20. S. J. Stein and L. Ugel, "Effect of Firing Conditions on Stability and Properties of Glaze Resistors," *Proc. 1968 Hybrid Microelectronics Symposium*, pp. 189–209, October 1968.
21. R. C. Musa, "Hybrid Packaging," *IEEE Workshop on Hybrid IC Technology*, pp. 6-1–6-4, March 1968.

22. G. K. Fehr, "Microcircuit Packaging and Assembly—State of the Art," *SST*, Vol. 13, No. 8, pp. 41–47, August 1970.

23. S. M. Stuhlbarg, "What is Needed to Get Started Packaging Hybrid Integrated Circuits," *1970 IEEE Conv.*, pp. 422–423, March 1970.

24. A. W. Postlethwaite, "Hermetic and Non-Hermetic Packaging," *SST*, Vol. 13, No. 8, pp. 67–70, 75, August 1970.

25. F. H. Bower, "Hermetic Sealing of Integrated Circuit Packages," *SST*, Vol. 13, No. 8, pp. 56–61, August 1970.

26. G. R. Cole, "Hermetic Seals by Thick Film Techniques," *SST*, Vol. 11, No. 8, pp. 43–46, August 1968.

27. R. C. Chalman, "Notes on Sealing Tekform Metal Packages and Flatpacks," Tekform Products Company, Sealing Notes.

28. R. Boggs and E. Kanazawa, "Packaging Complex Multi-Chip Systems—Design Considerations and Applications," *Proc. 1968 Hybrid Microelectronics Symposium*, pp. 1–10, October 1968.

29. A. S. Budnick, "Manufacturing the Plastic Dual In-Line Integrated Circuit," *SST*, Vol. 11, No. 8, pp. 37–42, August 1968.

30. R. J. Anderson et al , "Selection and Control of Plastics for Semiconductor Packaging," *Proc. 1968 Hybrid Microelectronics Symposium*, pp. 505–513, October 1968.

31. H. Hirsch, "Resin Systems for Encapsulation of Microelectronic Packages," *SST*, Vol. 13, No. 8, pp. 48–55, August 1970.

32. J. H. Seely and O. R. Gupta, "Thermal Conductance and Its Effect on Electronic Packages," *1970 International Hybrid Microelectronics Symposium*, pp. 4.7.1–4.7.11, November 1970.

33. F. H. Bower, "Hermetic Packages and Sealing Methods," *1970 International Hybrid Microelectronics Symposium*, pp. 6.2.1–6.2.10, November 1970.

34. M. Ferland, "Important Considerations in Selecting a Manual IC Tester," *SST*, Vol. 12, No. 3, pp. 49–54, March 1969.

35. G. D. Morant, "Automated Hybrid Circuit Testing," *Proc. 1968 Hybrid Microelectronics Symposium*, pp. 43-70, October 1968.

36. G. W. Smith, III, and L. E. Little, "Packaging of Precision Analog Microcircuits," *1970 International Hybrid Microelectronics Symposium*, pp. 4.1.4–4.1.7, November 1970.

37. R. R. Prudhomme, "Developing the Quality Assurance Requirements for a Custom Thin-Film Circuit Program," *SCP and SST*, Vol. 10, No. 8, pp. 58–61, August 1967.

Chapter 7

THE HYBRID MICROCIRCUIT VENDOR

This chapter considers three approaches to hybrid circuit development and manufacture: first, completing all design, prototyping, and production in-house; second, designing all circuits and conducting hybrid prototyping in-house but contracting full-scale production to a vendor; third, turning over all details of hybrid prototyping and production after the circuit design phase to a hybrid vendor.

The facilities of a typical production plant will be described and criteria for judging a prospective vendor will be listed. Technical information needed by the vendor before he can begin prototyping or production also is outlined. Finally, questions of delivery time and cost are discussed.

COMPLETE IN-HOUSE HYBRID
PRODUCTION FACILITY

Complete in-house hybrid design, prototyping, and production is the best way of maintaining confidentiality about proprietary designs and secretive end applications for the hybrid. Moreover, design engineers sometimes judge that the layout of critical circuits should be done in-house. Special quality control requirements may mandate an in-house capability. Although economic considerations usually strongly suggest turning the more advanced stages of prototyping and production over to a vendor (1), in circumstances where large numbers of hybrids are to be produced continuously, a complete in-house plant can be profitable.

ADVANTAGES OF A VENDOR

Cost and time savings usually justify employing the services of an outside vendor. A minimum in-house prototype facility can be built for $25,000 to $50,000 and a well-equipped facility for $100,000 (2), but a full-scale thick film production line will cost over a $\frac{1}{4}$ million. Moreover, continuing expenditures are required to keep this plant at the state-of-the-art in hybrid technology (3).

On the other hand, vendors can deliver prototype samples within three weeks to three months, depending on the complexity of the circuit required; in any case, vendors require less time than that needed to implement an in-house hybrid facility. The competition between microcircuit vendors for prototype business tends to keep costs competitive and also forces continued state-of-the-art facilities improvement. A competent circuit vendor usually has an engineering department that is knowledgeable about components that are critical to performance. This group is familiar with optimum types of active devices for given circuits, with design tradeoffs, and with package selections that insure compatibility between the hybrid and associated electronic systems.

The vendor builds microcircuits quickly and holds tooling costs to a minimum due to the large volume of hybrids he produces. Providing quick turnaround time in prototype manufacture, he can set up pilot runs, deliver circuits to his customer for tests, modify these circuits as needed, then quickly begin major production. An outside vendor can relieve the customer of a major quality control responsibility by guaranteeing to deliver quantities of hybrids tested and certified to pass all functional tests specified by the customer (4).

VENDOR DESIGN SERVICES

Besides manufacturing thick film modules, a vendor applies design knowledge and skills with the aid of computer facilities to assure correct, reliable, and economical design of a customer's hybrid circuit. Since he is familiar with the reliability and tolerances of passive, screened, or chip components, with thick film process limitations, with available transistor, diode, and IC chips, and with all available package configurations, the vendor can be of great assistance to his customer in the design phase of hybrid development.

Calling on an extensive library of dc, ac, and transient computer-aided design programs, the vendor can show interactions between circuit elements and identify the effects of part parameter variations on long-term circuit performance (5).

FACILITIES AND QUALITY CONTROL

Vendor-supplied hybrids usually are produced in large, well-lighted, air conditioned, air purified plants. A partial list of production equipment would include apparatus for the following:

- Slice scribing and separation.
- Slice storage (to minimize oxidation and moisture absorption).
- Automatic screen printing.
- Firing conductor, resistor, and dielectric pastes.
- Resistor trimming.
- Prestressing and preaging substrates.
- Photographing artwork.
- Rubylith cutting.
- Screen production.
- Bonding active and passive chips.
- Encapsulating or packaging finished circuits.

After initial design and packaging decisions are made a pilot line is initiated to provide hybrid circuit prototypes for early testing and to establish process control specifications and a quality control program to be used in later full-scale manufacture. Quality control includes material inspection, process control, and final functional, environmental, and life tests.

A complete quality control department would have access to the following test equipment:

- Transistor curve tracers.
- Temperature and environmental test chambers.
- Regulated power supplies.
- Signal generators.
- Oscilloscopes.
- Pulse generators.
- Digital voltmeters.
- Wheatstone and other bridges.
- Frequency counters.
- Decade resistor boxes.
- Standards.
- Crystal controlled time-base calibration oscillators.
- Hermetic seal testers.
- Lead wire pull and peel testers.
- Metallurgical microscopes.
- Shadow graphs.
- Strip chart recorders for furnace profiling and process control.
- Cermet viscosity meters.
- In-process visual and electrical probes.

NOTES ON SELECTING A VENDOR

As a minimum, one ought to choose a vendor who will remain active in hybrid production long enough to provide replacements and duplicate hybrids for later applications.

Ideally the vendor should have an engineering applications group sophisticated enough to understand the circuit design function requested, guaranteeing that his hybrid circuits will meet the black-box specifications of his customer. Further, it is reassuring to know that one's own vendor is producing an off-the-shelf hybrid similar to the circuit requested. Finally, a list of alternate suppliers should be compiled if possible (6).

Hybrid Microcircuit Design Data Sheet — Collins Radio Company

COLLINS

COMPANY _____

ENGINEER _____ DATE _____

ADDRESS _____

TELEPHONE _____

CIRCUIT REQUIREMENTS

1. Schematic (please attach as separate sheet)

2. Resistors

NUMBER	R1	R2	R3	R4	R5	R6	R7	R8	R9	R10
VALUE										
TOLERANCE										
CURRENT										
NUMBER	R11	R12	R13	R14	R15	R16	R17	R18	R19	R20
VALUE										
TOLERANCE										
CURRENT										

3. Capacitors

NUMBER	C1	C2	C3	C4	C5
VALUE					
TOLERANCE					

4. Inductors

NUMBER	L1	L2	L3	L4
VALUE				
TOLERANCE				

5. Semiconductors

NUMBER	Q1	Q2	Q3	Q4	Q5	Q6	Q7	Q8	Q9	Q10
TYPE										
NUMBER	Q11	Q12	Q13	Q14	Q15	Q16	Q17	Q18	Q19	Q20
TYPE										

PACKAGING REQUIREMENTS

1. Package size preferred: ⅜ inch by ⅜ inch ☐   ⅝ inch by ⅝ inch ☐   1 inch by 1 inch ☐   1 inch by 1 inch integral ☐

2. List any specially oriented external pin connectors

3. Special package marking

Figure 7.1.   Typical forms for providing design details to prospective hybrid vendors.   Courtesy of Collins Radio Company and Dickson Electronics Corp.

191

PERFORMANCE REQUIREMENTS (ELECTRICAL)

PERFORMANCE REQUIREMENTS (ENVIRONMENTAL)

Operating temperature range _____ °C to _____ °C

Non-operating temperature range _____ °C to _____ °C

Shock _____

Vibration _____

Acceleration _____

SPECIAL SCREENING REQUIREMENTS

OTHER REQUIREMENTS OR COMMENTS

QUANTITY AND DELIVERY REQUIREMENTS

Quotation required by (date) _____   Budgetary ☐   Firm ☐

Prototype required by (date) _____   Quantity _____

Production required by (date) _____   Total quantity _____ Rate per week

Send to Collins Radio Co. HYBRID CIRCUITS – M/S 407–201 Dallas, Texas 75207

Figure 7.1. Continued

193

HYBRID SPECIFICATIONS

Analog

A) INPUT CHARACTERISTICS

Frequency _____ Hz min. _____ Hz max.

Source Impedance _____ Ω

WAVEFORM

+ __V

0

− __V

_____ t _____ Sec./Div.

Indicate Maximum Positive and Negative Excursions

Digital

A) INPUT CHARACTERISTICS

Logic "1" _____ V min. _____ V max.

Logic "0" _____ V min. _____ V max.

tr _____ max. tf _____ max.

Pulse Width _____ min.

Rep. Rate _____ max.

Source Impedance _____ Ω

B) OUTPUT CHARACTERISTICS

Logic "1" _____ V min. _____ V max.

Logic "0" _____ V min. _____ V max.

tr _____ max. ts _____ ma.

max.

Load Impedance _____ max. ﬨ _____ max.

MISC. _____ Ω

C) TIMING DIAGRAM

Input

Output

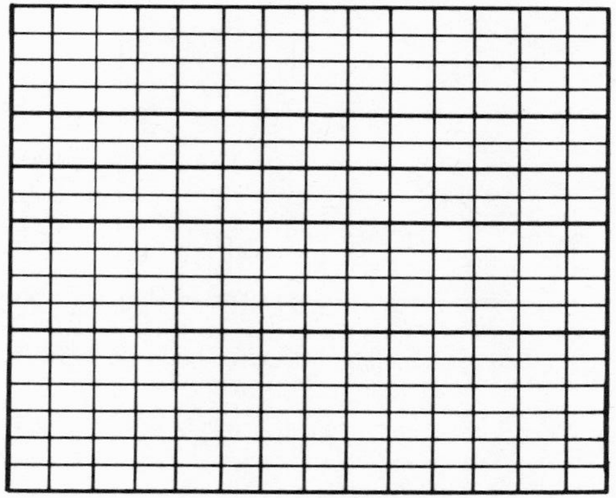

t _____ Sec./Div.

B) OUTPUT CHARACTERISTICS

Gain _____ min.

Frequency Response @ 3 dB _____ Hz

Load Impedance _____ Ω

WAVEFORM

+ __ v

0

− __ v

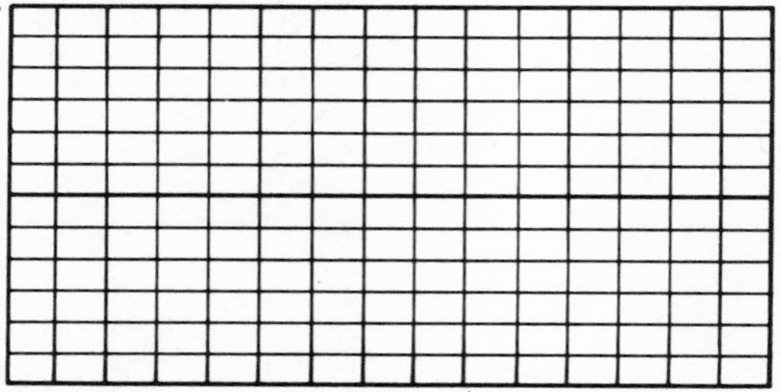

t _____ Sec./Div.

Indicate Maximum Positive and
Negative Excursions

195

Figure 7.1.   Continued

RESISTORS

	OHMS	% TOL	WATTS	TC
R				
R				
R				
R				

CAPACITORS

	CAP	% TOL	VOLTAGE	TC
C				
C				
C				
C				

POWER SUPPLY

	MIN	TYP	MAX	% RIP
Vc 1				
Vc 2				
Vc 3				
Vc 4				

SEMICONDUCTORS

PART NO.	JEDEC NO.	CHAR.	PART NO.	JEDEC NO.	CHAR.

MONOLITHIC IC'S

PART NO.	MFG. NO.	CHAR.	PART NO.	MFG. NO.	CHAR.

PACKAGE TYPES

STORAGE TEMP. RANGE _____ LOW _____ HIGH

OPERATING TEMP. RANGE _____ LOW _____ HIGH

_____ NO PREFERENCE

_____ TO-12   _____ CN48A (TO-8 mod)   _____ TO-86

_____ TO-78 (mod)   _____ CN48B (TO-8 mod)   _____ TO-87   _____ OTHER

_____ TO-97   _____ CN50 (TO-8 mod)   _____ FT-35

_____ TO-101   _____ TO-84   _____ TO-116

_____ NUMBER OF LEADS   SIZE RESTRICTIONS H _____ W _____ L _____

**ALL ADDITIONAL INFORMATION
SHOULD BE ATTACHED TO THIS FORM**

Figure 7.1. Continued

197

CHECK DESIRED TESTS

		MIL-STD-883 METHOD NO.
*	1.) Functional Test	
*	2.) Constant Acceleration	2001
*	3.) Gross Leak	1002
*	4.) Fine Leak Helium (10^{-8} cc/sec.)	1014
*	5.) Temperature Cycle ($-55°C$ to $85°C$ 10 cycles) . . .	1010
*	6.) High Temperature Storage ($200°C$ for 48 hours) . . .	1008
	No Further Testing Required.	
	7.) Barometric Pressure	1001
	8.) Moisture Resistance	1004
	9.) Steady State Operating Life	1005

_____ 10.) Salt Atmosphere (D) 1009

_____ 11.) Thermal Shock 1011

_____ 12.) Mechanical Shock 2002

_____ 13.) Solderability (D) 2003

_____ 14.) Lead Integrity (D) 2004
     A. Lead Fatigue
     B. Lead Tension

_____ 15.) Vibration
     A. Fatigue 2005
     B. Variable 2007

_____ Environmental Specification Prints Enclosed

_____ Other _____

_____

_____

(D) Destructive Type Tests

✱ These tests are performed as standard procedure.

Figure 7.1. Continued

199

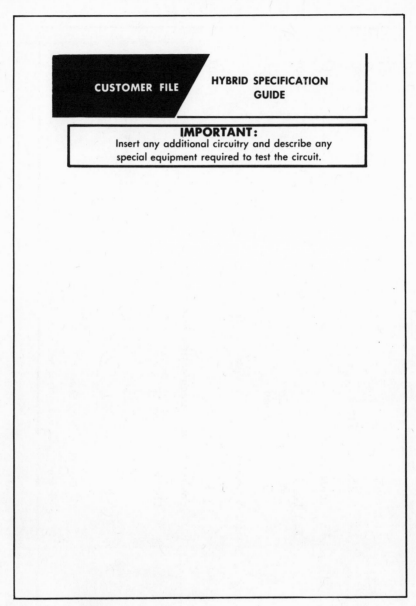

CUSTOMER FILE

HYBRID SPECIFICATION GUIDE

IMPORTANT:
Insert any additional circuitry and describe any special equipment required to test the circuit.

Figure 7.1. Continued

COMPANY: _____

ADDRESS: _____

CITY _____ STATE _____ ZIP _____

DIVISION &/OR DEPT.: _____

MAIL STATION: _____

TELEPHONE: _____ EXT.: _____

CUSTOMER CONTACT (NAME): _____

TITLE: _____

TYPE OF CIRCUIT (FUNCTION): _____

_____

DRAWING NUMBERS

Figure 7.1. Continued

CUSTOM HYBRID PROCUREMENT

It is always advantageous to begin discussions with a hybrid vendor at the earliest possible stage of systems design. In presenting black-box specifications to the vendor, consider whether he can substitute an off-the-shelf item with little or no modification, thus saving time and money.

A vendor inquiry should include the following: (1) a detailed circuit schematic of a device that has been breadboarded and tested and is known to function properly; (2) a black-box input-output parameter schedule; (3) a package configuration preference; (4) a note on the maximum size allowable for the module; (5) a survey of environmental details including operating temperature range, humidity, corrosive atmosphere if present, shock, vibration, and the need for hermetic sealing; (6) a statement of the nature of the application—whether military, industrial, or commercial; (7) a list of active device specifications containing only parameters of real importance so that the hybrid vendor can use devices that are readily available to him and thus deliver a microcircuit in the shortest possible time; (8) a list of values, tolerances, temperature coefficients, and other essential characteristics for all resistors and capacitors; (9) a table of values of tolerances, Q's, and dimensions of all inductors and transformers with the vendor given the option, if possible, of designing these out of the circuit; (10) a desired production schedule including delivery dates, quantities, and cost objectives; and (11) an outline of reliability and quality control requirements detailing essential tests. Figure 7.1 is a typical design data sheet for supplying the foregoing information to a vendor.

The interests of both customer and vendor are served best by bringing the vendor into the design process as early as possible, by allowing him a wide latitude in the selection of active devices and inductors, by thoroughly completing designing and breadboarding before initiating hybrid layout, and by discussing with the vendor as fully as possible the application for which the hybrid is intended.

CIRCUIT COSTS AND DELIVERY TIMES

The price of a vendor-supplied hybrid circuit depends on its complexity, difficulty of manufacture, design involvement of the hybrid vendor, parts cost, and quality control and reliability requirements.

Considering complexity versus cost, a simple passive network may have a per unit price of 30¢ to $4.00, depending on component values, package requirements, and other specifications. A more complicated audio or RF oscillator may sell for $10.00. But a 20-W, 2-GHz microwave amplifier might cost more than $1000. One-time engineering charges can be as little as $100 or as much as $20,000, depending on the number of man-hours of engineering time needed. Single-function chips for commercial applications usually cost the customer $1 to $4 per unit. In military applications tight specifications, complicated circuits, and strict reliability demands send per unit prices into the hundreds of dollars (7–9).

Prototypes generally can be delivered within two weeks to three months after approval of circuit layout drawings. Large-quantity production deliveries are possible within two to three months. The availability of semiconductors determines the speed of delivery.

SUMMARY

A comparison was made between the cost of producing circuits in-house or by an outside vendor. Economic realities usually strongly suggest the latter; certain situations, however, mandate complete in-house production. Vendor design services, facilities, and typical quality control programs were discussed. The characteristics to be searched for in a vendor were mentioned and the information required before vendor production can be begun was listed. A composite of circuit costs and delivery times also was presented.

REFERENCES

1. W. B. Schlak, Jr. (Ledex Microelectronics), private communication, August 1970.
2. J. D. Dale, "Experience of a User in Getting Started with Thick Film," *IEEE Conv., Technical Applications Session*, TA-1.9, p. 424, March 1970.
3. J. M. Cohen, "Getting Started with a HIC Vendor," *IEEE Conv., Technical Applications Session*, TA-1.1, p. 410, March 1970.
4. "Hybrid Microcircuit Design Manual for the '70's," Circuit Technology, Inc., p. 5, 1970.
5. W. L. Roberts (Collins Radio), private communication, August 1970.
6. Cohen, *op. cit.*, p. 410.
7. J. R. M. Alger (General Electric), private communication, September 1970.
8. R. Ho (American Electronic Laboratories), private communication, August 1970.
9. Roberts, *op. cit.*

APPENDIX

ADDITIONAL INFORMATION

The following list of books, journals, and companies concerned with hybrid technology is provided as an aid to readers seeking further information on thick film topics. Unfortunately, space limitations have required the omission of many excellent sources from this list. We regret this and the errors that cannot be completely avoided in such a compilation. Appearance here should in no case be considered as either an endorsement or a recommendation by the author or by the IEEE.

TEXTBOOKS AND CONFERENCE PROCEEDINGS

Proceedings of the IEEE Electronic Components Conferences, Institute of Electrical and Electronics Engineers, New York.
Proceedings of ISHM Symposia (1967–1972), ISHM Headquarters, Park Ridge, Ill.
Screen Printing of Microcircuits, Daniel C. Hughes, Jr., Somerville, N.J., Presco, Incorporated, 1967.
Thick Film Hybrid Microcircuit Technology, D. W. Hamer and J. V. Biggers, John Wiley and Sons, New York, 1972.
The Thick Film Microcircuitry Handbook, E. I. DuPont de Nemours and Co., Inc., Wilmington, Delaware.
Thick Film Microelectronics: Fabrication, Design and Application, M. L.

Topfer, Van Nostrand Reinhold, New York, 1971.
Thick Films Technology and Chip Joining, L. F. Miller, Gordon and Breach, New York, 1972.

JOURNALS AND PERIODICALS

EDN, Rogers Publishing Co., Inc., a subsidiary of Cahners Publishing Co., 270 St. Paul, Denver, Colo. 80206.
Electronic Engineering, Morgan Grampian Ltd., 28 Essex, London WC2.
Electronic Products, United Technical Publications, Inc., Division of Cox Broadcasting Corp., Garden City, N. Y. 11530.
Hybrid Microelectronics Review, P.O. Box 11685, Philadelphia, Pa. 19116.
IEEE Transactions on Parts, Hybrids and Packaging, Institute of Electrical and Electronics Engineers, 345 East 47th Street, New York, N.Y. 10017.
Proceedings of the IEEE, Institute of Electrical and Electronics Engineers, 345 East 47th Street, New York, N. Y. 10017.
Solid State Technology, Cowan Publishing Corp., 14 Vanderventer Ave., Port Washington, N. Y. 11050.
State of the Art Digest, State of the Art Inc., 1315 S. Allen Street, State College, Pa. 16801.
The Electronic Engineer, Chilton Company, Chestnut and 56th Street, Philadelphia, Pa. 19139.

Substrates

American Lava Corp.
Cherokee Blvd.
Chattanooga, Tenn. 37405
615-265-3411

Brush Beryllium Co.
St. Clair Ave.
Cleveland, O. 44110
216-486-4200

Centralab Division
Globe-Union, Inc.
5757 North Green Bay Ave.
Milwaukee, Wisc. 53201
414-228-1200

Coors Porcelain Co.
600 Ninth St.
Golden, Colo. 80401
303-279-6565

Erie Technological Products,
  Inc.
644 W. 12th St.
Erie, Pa. 16512
814-453-5611

National Beryllia Corp.
Greenwood Ave.
Haskell, N. J. 07420
201-839-1600

NGK Spark Plugs (USA), Inc.
NTK Special Ceramics Division
4010 Sawtelle Blvd.
Los Angeles, Calif. 90066
213-397-8184

Van Keuren Company
176 Waltham St.
Watertown, Mass. 02172

Varadyne, Inc.
1805 Colorado Ave. ,
Santa Monica, Calif. 90404
213-394-0271

Pastes

E. I. DuPont De Nemours and
  Co., Inc.
Electrochemicals Department
Electronic Products Division
Wilmington, Del. 19898

Electro-Science Laboratories,
  Inc.
1133 Arch St.
Philadelphia, Pa. 19107
215-563-2215

Bala Electronics Corp.
Cermet Division
14 Fayette St.
Conshohocken, Pa. 19428
215-828-4650

Owens-Illinois
Electronic Materials
Consumer and Technical
  Products Division
Toledo, O. 43601
419-242-6543

Resistor Trimmers

Apollo Lasers, Inc.
6365 Arizona Circle
Los Angeles, Calif. 90045
213-776-3343

Arvin Systems, Inc.
1482 Stanley Ave.
Dayton, O. 45404
513-222-8379

Axion Corp.
6 Commerce Park
Danbury, Conn. 06810
203-743-9281

Data Systems Corp.
301 East Airy St.
Norristown, Pa. 19401
215-277-4555

de Haart, Inc.
12 Wilmington Rd.
Burlington, Md. 01803

Hughes Aircraft Co.
3100 W. Lomita Blvd.
Torrance, Calif. 90509
213-534-2121

MPM Corp.
9 Harvey St.
Cambridge, Mass. 02140
617-876-7111

Mechanization Associates
140 South Whisman Road
Mountain View, Calif. 94040
415-967-4262

Micronetic, Inc.
204 Arsenal St.
Watertown, Md. 02172
617-926-2570

Raytheon Company
Microwave and Power Tube
  Division
Waltham, Md. 02154
617-899-8400

Spacerays, Inc.
Northwest Industrial Park
Burlington, Md. 01803
617-272-6220

TRW/Instruments
139 Illinois St.
El Segundo, Calif. 90245
213-535-0854

Union Carbide Corp.
Korad Department, Electronics
  Division
2520 Colorado Avenue
Santa Monica, Calif. 90406
213-393-6737

S. S. White Co.
Division of Pennwalt Corp.
201 East 42nd St.
New York, N. Y. 10017
212-661-3320

Resistor Chips

Airco Speer
800 Theresa St.
St. Marys, Pa. 15857
814-834-2801

American Components, Inc.
8th Ave. and Harry St.
Conshohocken, Pa. 19428
215-828-6240

Dale Electronics, Inc.
28th Ave.
Columbus, Neb. 68601
402-564-3131

Dickson Electronics Corp.
P. O. Box 1390
Scottsdale, Ariz. 85252
602-947-2231

EMC Technology, Inc.
1300 Arch St.
Philadelphia, Pa. 19107
215-563-1340

Semiconductor Products Corp.
Kulite
1038 Hoyt Ave.
Ridgefield, N. J. 07657
201-945-3000

Motorola, Inc.
Semiconductor Products
  Division
Phoenix, Ariz. 85036
602-273-6900

Solid State Scientific, Inc.
Montgomeryville, Pa. 18936
215-855-8400

Sprague Electric Co.
Marshall St.
North Adams, Md. 01247
413-664-4411

United Aircraft Corp.
Electronic Components Division
4850 Street Rd.
Trevose, Pa. 19047
215-355-5000

Varadyne, Inc.
1805 Colorado Ave.
Santa Monica, Calif. 90404
213-394-0271

Capacitor Chips

Aerovox Corp.
740 Bellevile Ave.
New Bedford, Md. 02741
617-994-9661

American Components, Inc.
8th Ave. and Harry St.
Conshohocken, Pa. 19428
215-828-6240

American Lava Corp.
Cherokee Blvd.
Chattanooga, Tenn. 37405
615-265-3411

Centralab Division
Globe-Union, Inc.
5757 N. Green Bay Ave.
Milwaukee, Wisc. 53201
414-228-1200

Dickson Electronics Corp.
P. O. Box 1390
Scottsdale, Ariz. 85252
602-947-2231

Erie Technological Products,
Inc.
644 West 12th St.
Erie, Pa. 16512
814-453-5611

Monolithic Dielectrics, Inc.
P. O. Box 647
Burbank, Calif. 91503
213-848-4465

Motorola, Inc.
Semiconductor Products Division
Phoenix, Ariz. 85036
602-273-6900

Sprague Electric Co.
Marshall St.
North Adams, Md. 01246
413-664-4411

United Aircraft Corp.
Electronic Components Division
4850 Street Rd.
Trevose, Pa. 19047
215-355-5000

U. S. Capacitor Corp.
2151 North Lincoln St.
Burbank, Calif. 91504
213-843-4222

Varadyne, Inc.
1805 Colorado Ave.
Santa Monica, Calif. 90404
213-394-0271

Vitramon, Inc.
P. O. Box 544
Bridgeport, Conn. 06601
203-268-6261

Motorola, Inc.
Semiconductor Products Division
Phoenix, Ariz. 85036
602-273-6900

Bonders

Hugle Industries, Inc.
625 North Pastoria Ave.
Sunnyvale, Calif. 94086
408-738-1700

Kulicke and Soffa Industries
135 Commerce Dr.
Fort Washington, Pa. 19034
215-646-5800

Precision Equipment Co., Inc.
1246 Central Ave.
Hillside, N. J. 07205
201-351-4442

Tempress Industries
980 University Ave.
Los Gatos, Calif. 95030
408-356-8151

Unitek Corp.
Weldmatic Division
1820 South Myrtle Ave.
Monrovia, Calif. 91016
213-359-8367

Chip Inductors

Piconics
Cummings Rd.
Tyngsboro, Md. 01879
617-649-7501

Active Devices

Amperex Electronic Corp.
Semiconductor Division
Slatersville, R. I. 02876
401-762-9000

Dickson Electronics Corp.
P. O. Box 1390
Scottsdale, Ariz. 85252
602-947-2231

Dionics, Inc.
65 Rushmore St.
Westbury, N. Y. 11590
516-997-7474

Fairchild Semiconductor
Mountain View, Calif. 94040
415-962-5011

Intersil
10900 North Tantau Ave.
Cupertino, Calif. 95014
408-257-5450

Motorola, Inc.
Semiconductor Products Division
Phoenix, Ariz. 85036
602-273-6900

National Semiconductor
2900 Semiconductor Dr.
Santa Clara, Calif. 95051
408-732-5000

Raytheon Co.
350 Ellis St.
Mountain View, Calif. 94040
415-968-9211

Solid State Scientific, Inc.
Montgomeryville, Pa. 18936
215-855-8400

Texas Instruments, Inc.
P. O. Box 512
Dallas, Texas 75222
214-238-2011

United Aircraft Corp.
Electronic Components Division
4850 Street Rd.
Trevose, Pa. 19047
215-355-5000

Screening Supplies

Affiliated Manufacturers, Inc.
U. S. Route 22
Whitehouse, N. J. 08888
201-534-2103

Aremco Products, Inc.
P. O. Box 145
Briarcliff Manor, N. Y. 10510
914-762-0685

Buckbee Mears Co.
245 East 6th St.
St. Paul, Minn. 55101
612-227-6371

Graining Equipment Co.
Nashville, Tenn. 37203
615-255-4179

Pelmore Laboratories, Inc.
401 Lafayette St.
Newtown, Pa. 18940
215-968-3825

Towne Laboratories, Inc.
1 U. S. Highway 206
Somerville, N. J. 08876
201-772-9500

Ulano
210 East 86th St.
New York, N. Y. 10028
212-628-7960

Printing Equipment

Affiliated Manufacturers, Inc.
U. S. Route 22
Whitehouse, N. J. 08888
201-534-2103

Aremco Products, Inc.
P. O. Box 145
Briarcliff Manor, N. Y. 10510
914-762-0685

de Haart, Inc.
12 Wilmington Rd.
Burlington, Md. 01803
617-272-0794

Engineered Technical Products
Box 1465
Plainfield, N. J. 07061
201-756-2160

Jos. E. Podgor Co.
Box 1714
Philadelphia, Pa. 19105
215-WA5-7878

Precision Systems Co., Inc.
U. S. Route 22, Box 148
Somerville, N. J. 08876
201-722-7100

Furnaces

BTU Engineering Corp.
Bear Hill Rd.
Waltham, Mass. 02154
617-894-6050

C. I. Hayes, Inc.
825 Wellington Ave.
Cranston, R. I. 02910
401-461-3400

W. P. Keith Co., Inc.
8323 Loch Lomond Dr.
Pico Rivera, Calif. 90660
213-723-1375

Trent, Inc.
201 Leverington Ave.
Philadelphia, Pa. 19127
215-482-5000

Watkins-Johnson Co.
3333 Hillview Ave.
Palo Alto, Calif. 94304
415-326-8830

Packages

American Lava Corp.
Cherokee Blvd.
Chattanooga, Tenn. 37405
615-265-3411

Coors Porcelain Co.
600 Ninth St.
Golden, Colo. 80401
303-279-6565

Sprague Electric Co.
Semiconductor Division
Concord, N. H. 03301
603-224-1961

Tekform Products Co.
2780 Coronado St.
Anaheim, Calif. 92806
714-630-2340

VENDORS

American Electronic
   Laboratories, Inc.
Richardson Rd.
Colmar, Pa. 18915
215-822-2929

Amperex Electronic Corp.
99 Bald Hill Rd.
Cranston, R. I. 02910

Cermex Corp.
695 Broadway
Westwood, N. J.

Circuit Technology, Inc.
160 Smith St.
Farmingdale, N. Y. 11735
516-293-8686

Collins Radio Company
Dallas, Tex. 75207
214-235-9511

Film Microelectronics, Inc.
17A St., Highland Industrial
   Park
Burlington, Mass. 01803
617-272-5650

General Electric Co.
Hybrid Operations
Electronic Components Division
Syracuse, N. Y. 13201
315-456-0123

General Instrument Corp.
Integrated Circuits Division
Hicksville, N. Y. 11802
516-733-3243

Globe-Union, Inc.
Centralab Semiconductor
   Division
El Monte, Calif. 91734

Lansdale Microelectronics, Inc.
Colmar, Pa. 18915
215-822-0155

Ledex Microelectronics
123 Webster St.
Dayton, O. 45402
513-222-6992

United Aircraft
Electronic Components Division
Trevose, Pa. 19047
215-355-5000

INDEX